John W Urquhart

Electro-motors

A Treatise on the Means and Apparatus Employed in the Transmission...

John W Urquhart

Electro-motors
A Treatise on the Means and Apparatus Employed in the Transmission...

ISBN/EAN: 9783337185947

Printed in Europe, USA, Canada, Australia, Japan

Cover: Foto ©berggeist007 / pixelio.de

More available books at **www.hansebooks.com**

ELECTRO-MOTORS:

A Treatise

ON THE MEANS AND APPARATUS EMPLOYED IN THE
TRANSMISSION OF ELECTRICAL ENERGY AND
ITS CONVERSION INTO MOTIVE POWER.

FOR THE USE OF ENGINEERS AND OTHERS.

By J. W. URQUHART, Electrician.

Author of "Electro-Plating: a Practical Handbook," "Electric Light,"
"Electro-Typing: a Practical Manual," etc.

WITH NUMEROUS ILLUSTRATIONS.

Manchester:
WILLIAM T. EMMOTT, BLACKFRIARS STREET.
London:
TRÜBNER & CO., LUDGATE HILL.
1882.

PREFACE.

This work is intended to convey, to engineers and others interested in the subject, an explanation, in conveniently plain terms, of the leading electrical and magnetic principles involved in the transmission of electrical energy and its subsequent conversion into motive power. It also gives examples of the means and apparatus employed in the working of electric railways, and other instances of the electrical translation of power. It is intended further to prepare the way towards a more thorough study of the correlative links between latent or potential energy, electricity, magnetism, and actual or active energy.

The mere question of obtaining motive power from electricity is neither new nor startling; but it is only since the discovery of the means of economically producing powerful currents of electricity that motion by this method became practicable for general purposes and the propulsion of railway vehicles. When currents of electricity were first utilised for practical purposes, and for many years afterwards, the cost of generating them precluded their application to the production of motion. They were only used for telegraphic signalling and for electro-metallurgical purposes. By the discoveries and developments of the past few years the whole aspect of the question has undergone a great change. The most powerful currents were, by the introduction of the dynamo-electric principle, fourteen years ago, placed at our disposal. The cost of producing these currents was immensely diminished. They could be evolved from the energy of coal in the steam engine, and they could be conveyed to a distance, to be then re-converted into mechanical motion, light, heat, or other required form of active energy. The waste powers of nature, such as waterfalls, could be brought into requisition for the development of powerful currents, and it at once became practicable to produce motion or light by means of these currents. The first might be employed in the propulsion of railway trains, and in moving machinery; the second might be used in the lighting of the highways. The rate of progress so far has been rapid, but more attention was at first paid to electric lighting than to the transmission or distribution of power. Now, however, that several instances of the application of the latter principle exist, in the form of electric railways and otherwise, its utilisation is only a question of time. The first permanent electric railway at Berlin has proved so satisfactory in its working that it has been considerably extended. The first section was from Berlin to Lichterfelde, and it has since been extended successively to Tetlow and Potsdam, the further extension to Steglitz being at present in course of construction. The German Government have further granted a concession for the formation of an electric railway from Eisenach to Wartburg.

The present volume may therefore be considered as an attempt not only to elucidate the fundamental principles underlying these means of translating power, but to explain the construction of the apparatus employed, with particulars of examples of what has already been accomplished in this direction. It appears necessary to explain that the construction of dynamo-electric machines is only cursorily treated, since it has already received attention in works devoted to electric lighting.

The author is glad to have this opportunity to tender his thanks for the liberal encouragement accorded to his previous endeavours in the domain of electro-metallurgy and electric lighting. He has to acknowledge his indebtedness for valuable assistance and interesting facts in the present instance to Messrs. Siemens Brothers and Co., of London; MM. Trouvé, Fontaine, and Deprez, of Paris; Sir William Armstrong, F.R.S.; and other pioneers of progress in electrical science.

London, January, 1882.

CONTENTS.

CHAPTER I.
Introduction . . . . 1 PAGE

CHAPTER II.
On the Dynamical Nature of Electric Currents . . 14

CHAPTER III.
Electrical Measurements 36

CHAPTER IV.
Electro-Magnetic Force 48

CHAPTER V.
Electro-Magnets and Armatures 53

CHAPTER VI.
Electric Accumulators or Magazines 67

CHAPTER VII.

The Construction and Efficiency of Electro-Motive Machines 78

CHAPTER VIII.

Electric Railways 121

CHAPTER IX.

Minor Applications of Electro-Motive Energy . . . 139

CHAPTER X.

Fragmentary Information 173

INDEX.

	PAGE
Amalgamation of Zinc	176
Annealing	55
Armature, Siemens'	95
,, Iron	57
,, Gramme's	113
Ascertaining Direction of Electric Currents	22
Attraction	92
Circuit	21
Collectors or Brushes	88
Coercive Force	177
Construction and Efficiency of Electro-Motors	78
Conducting Wires	60
Conductors and Insulators	19
Commutator Brushes	88
Dynamo and Magneto Machines, Distinction between ...	84
Dynamo-Electric Apparatus as Current Generators ...	86
Dynamical Nature of Electrical Currents	14
Determination of Electro-Motive Force	46
Dentists' Drills, Electric	146
Economical Working	171
Efficiency of Electro-Magnetic Apparatus	65
Electro-Magnetic Field	93
Electric Pen	148

INDEX.

	PAGE
Electro-Motors for Sewing Machines	151
Electric Railways	121
,, ,, Expenditure for	130
Electric Routing Machines	140
Elements, Size of	170
Electro-Motive Force	15
Electric Accumulators or Magazines	67
Electro-Magnetic Cores	57
,, Force	48
,, Solenoids	60
Electric Poles	21
Electro-Magnets and Armatures	53
Electro-Motive Force, Inverse	117
Electrical Measurements	36
Electricity, Minor Sources of	155
Electric Stone-Engraving Apparatus	143
Electro-Motive Energy, Minor Applications of	139
Extension of the Poles	59
Force, Coercive	177
Fragmentary Information	173
Froment's Model	94
Gramme's Armature	113
Grouping Cells	169
Illustrations of the Dynamical Effects of Electrical Currents	22
Induction	24
Instruments used in Electrical Measurements	37
Internal Resistance of Magneto-Electric Inductors	35
Inverse Electro-Motive Force	117
List of Insulated Wires	61
Local Action	176
Machines connected in Parallel Circuit	85

INDEX. xi

	PAGE
Magnetisation	56
Magnetic Induction	26
„ Attraction	50
Magnetisation, Relative Intensities of	178
Magnetic Tension and Electro-Motive Force	34
Measurements of Resistance	43
Modified Armature, Siemens'	111
Model, Froment's	94
Motors, Reciprocating	116
Obach's Tangent Galvanometer	38
Postal Messages, Electric-Transmitter for	153
Precautions against Short-Circuiting	88
Principles of Motion in Electro-Motors	89
Production of Magneto-Electric Currents	31
„ Mechanical Motion	25
Railways, Electric	121
Reciprocating Motors	116
Relative Intensities of Magnetisation	178
Resistance of the Conductor	16
„ Measurement of	43
Residual Magnetism	52
Resistance Coils and the Rheostat	43
Rotating Engines for Vacuum Tubes	148
Routing Machine, Electric	140
Sewing Machines, Electro-Motors for	151
Short-Circuiting, Precautions against	88
Shunts	42
Siemens' Armature	95
„ Modified Armature	111
Size of the Elements	170
Small Currents, Working Cost of	173
Soft Iron	54

	PAGE
Source of Electricity for Experiments	20
Steel	55
Stone-Engraving Machine, Electric	143
Tempering	55
Transmitter for Postal Messages	153
Vacuum Tubes, Engine for	148
Voltaic or Zinc Generators	159
Wires, list of insulated	61
Working Cost of Small Currents	173
Zinc, amalgamation of	176

ELECTRO-MOTORS.

CHAPTER I.

Introductory.

THE powerful attractive influence exerted by soft iron, around which an electric current is caused to circulate, drew the attention of practical men, at an early period in the history of electrical science, towards the conception of a moving force derived from some such means.

When it became known that continuous currents of electricity were obtainable by the combustion of zinc in the voltaic cell, further curiosity was aroused as to the practicability of obtaining an economical motive power from this mysterious agency by means of electro-magnetism. Various experimenters produced electro-magnets capable of exerting an attractive force of some hundred-weights near to their polar extremities. It was early discovered that the immense attractive influence so conferred upon the iron might be instantly destroyed by simply suspending the flow of the current, and that it might be as rapidly revived by restoring it. It was discovered that when two electro-magnets were caused to attract each other, they might afterwards be incited to mutually repel, by merely reversing the course of the current in one of them.

The consideration of these facts led many persons to conceive an electro-motor through which an electric current would pass; and it was not difficult to believe that such a machine might prove more safe and less troublesome than a steam engine for purposes of moving machinery and propelling ships.

About the year 1834 a Russian philosopher, Professor Jacobi, began to devote his attention, at the instigation of the Russian Government, to the question of obtaining an economical moving force from electricity. Applying the then known means of converting electric into mechanical force, he succeeded in constructing an engine, the motive power of which was derived from an electric current, maintained by one of the first forms of voltaic battery. This machine was probably the first electro-motor ever constructed.

Encouraged by partial success, Jacobi continued his experiments, and in the year 1838 succeeded in constructing an electric engine capable of propelling a boat 28ft. in length, containing ten persons. The vessel moved at four or five miles an hour on the Neva. This machine was actuated by the electric current evolved by a battery of sixty-four cells, which, like most other batteries, converted the potential energy of zinc into an electric current. The boat was moved through the medium of paddle wheels. The same form of motor was subsequently applied by its inventor to the driving of machinery, but the great cost of the zinc consumed in the electric battery necessitated the abandonment of the scheme.

A similar machine, in the form of a locomotive, was tried by Mr. Davidson, on the Edinburgh and Glasgow Railway, in 1832. It was 16ft. in length, weighed five tons, and travelled at about four miles an hour. An experiment similar to that made by Professor Jacobi was exhibited by Mr. Llewelyn, to the members of the British Association at Swansea, in the year 1848. The electro-motor used on that occasion was a great improvement upon those previously invented, but it also was found too wasteful of zinc in the generator to compete with the most inferior steam motors in point of working cost.

An infinite variety of electro-motors, embodying examples of the highest inventive powers, ingenuity, and constructive skill, was produced in the period which elapsed between Jacobi's invention and the publication of Dr. Joule's important investigation of the mechanical equivalent of heat. None of these engines could, however, convert electricity into mechanical motion so as to supersede steam engines in point of cheapness.

Dr. Joule's investigation had special reference to the question of the electro-motor, but it also proved many points not directly affecting these machines. It was undertaken chiefly in order to determine the relative potential energies of zinc and coal. By these means it was hoped that the question of the electro-motor might, so far as currents from zinc were concerned, be definitely settled. At that time zinc was the cheapest material by the consumption of which electricity could be evolved. The heat produced by the combustion of a pound of zinc could be ascertained, and it was equally easy to determine that due to the combustion of a pound of coal; but, although an experiment of this nature proved that a pound of coal yields about seven times the heat produced by the consumption of a similar weight of zinc, the fact was of little apparent significance until Joule showed, in a conclusive way, that *heat* has a definite value in relation to *work*—that, in short, a given amount of heat, no matter how obtained, was invariably equivalent to a corresponding amount of mechanical energy.

The energy or work power potential in zinc was proved to be equivalent to the heat yielded by that metal in the process of combustion, and the same was shown to be true of coal. Hence, it was comparatively easy to determine the relative costs of working coal and zinc motors when the ratio of their efficiencies was determined. The same masterly scientific investigation of this great question led to the conclusion that a coal motor was, scientifically considered, much less effective than a zinc or electro motor. But in relation to the question of working cost, the fact that coal possessed seven times the potential energy of zinc was of the greater importance. From this consideration it is evident that, as zinc is dearer than coal, the cost of energy derived from zinc must be very considerably greater than that derived from coal. With regard to this point, Professor Jenkin, F.R.S., remarks as follows, in his "Electricity and Magnetism" (p. 295): "In estimating that a zinc motor may be only fifty

INTRODUCTION.

times as dear as a coal motor, I assume that the electro-magnetic engine may be four times as efficient as the heat engine in transforming potential into actual energy."

These considerations show very clearly the main facts at issue in the case of transmuting the potential energy of zinc into current electricity, and thence into mechanical effect. They show conclusively that it is hopeless to expect that electro-motors, the power of which was derived from zinc, could compete with steam engines, and they thus exhibit at a glance the main cause of failure in the earlier types of the electro-motive machine. Nevertheless, the great convenience attending the use of small moving powers, conveyed through thin flexible conductors, may in many instances counterbalance the cost of working, and render *miniature* electro-motive apparatus, actuated by currents derived from the consumption of zinc, extremely useful.

Moreover, the efficiency of the older forms of electro-motor was very small. It is doubtful whether a return greater than 30 per cent. as mechanical force was available.

The investigation of the mechanical equivalent of heat brought to light other facts than those which led to the conclusion that zinc motors must be fifty times as expensive in working as coal motors. It showed that every form of energy was convertible into every other form. Thus, since it is evident that we can never create energy of any kind, we must confine ourselves to converting one form into another form. In the case of electric energy being transformed into mechanical effect, the most direct and effective connecting link which has yet been discovered is magnetism, or magnetic attraction and repulsion. It arises from the nature of matter that there must be a sensible loss of energy in the process of conversion. This loss appears chiefly as heat in the case of electricity, and as heat it is dissipated in the air. From this consideration it is evident that in transmuting any form of energy potential into actual motion through the agency of electricity, the main consideration must be to reduce the loss in the process to the smallest possible amount—we seek, in short, to obtain the maximum of effect from the minimum of expenditure.

Heat is convertible into electricity direct in small quantities, and we have thus a thermo-electric battery or converter, capable of furnishing a constant current, or, in other words, of converting a supply of heat into electricity direct. This current of electric energy is convertible into heat again, or into magnetism, thence to motive force through attraction and repulsion.

The dissolution of zinc, when it does not evolve heat, usually gives rise to electricity. In the voltaic cell the potential energy of zinc is frequently converted into heat when the conditions for the existence of current electricity are not provided; but as soon as the necessary conductive path is completed, a current of electricity flows, and the heat diminishes in direct proportion. In the electric circuit this current is convertible into a great many different forms of work.

Mechanical energy or motion is in turn converted into electric energy and magnetism. Upon this principle is based the dynamo-electric machine, by means of which any given moving force may be converted into an electric current with insignificant loss. This principle may be said to be at the root of the wonderful progress which has taken place within the past twelve years in the application of electricity to lighting, the propulsion of railway carriages, and other practical purposes. It has been said that heat is convertible directly into electricity. No method has yet been discovered by means of which this can be conveniently done on a large scale. But meantime the steam engine furnishes the most direct medium by which the potentiality of coal may be transmuted into electric currents. Other sources of motion are available, as the power of falling water and wind.

The powerful currents thus obtained, at a cost of less than 10 per cent. of the prime energy lost in conversion, may, as above mentioned, be again transformed into motive power. It is not difficult to perceive that by these means motive power, in the shape of an electric current, may be transmitted from one point to another through a suitable conductor. Thus arises the great question of the transmission of power. From one central point motive power may be distributed to numerous surrounding electro-motors. Thus the numerous small steam engines used in towns may be replaced by electro-motors of any required power, demanding no attention, and perfectly safe. One stationary condensing compound engine may by these means be employed to distribute its energy to a large number of electro-motors, so that the energy supplied may actually cost much less than the working of small steam engines on the spot.

Once the recently-developed dynamo-electric converters are regarded in this light there is scarcely a limit to the great facilities for utilising and transmitting motive power thus presented to the mind. Eminent authorities (Dr. C. W. Siemens, and others) have given it as their opinion that 1,000 horse power might by these means be conveyed for a distance of thirty miles, in a copper conductor of comparative small diameter.

The power of waterfalls may be utilised by means of this principle of transmission of power. Instances exist of dwelling-houses being supplied with electric light at night, and power by day, by utilising waterfalls. A familiar instance is presented in England, at Craigside, the residence of Sir William Armstrong, near Newcastle. The following particulars of this example, kindly furnished by Sir William Armstrong himself for this work, may be of interest :—

A waterfall of six horse power exists at a distance of 1,500 yards from the residence. This is caused to actuate a turbine, which is connected by a belt to a dynamo-electric converter, capable of transmuting about five horse power into a current of electricity. The current thus obtained is conveyed by a suitable conductor, buried in the ground, to the dwelling-house, where it is transformed into light, about 40 Swan electric lamps being used for that purpose. By day, the current is turned into an electro-motor, in the form of a second dynamo-electric machine, at a distance of 800 yards from the waterfall, and this electro-motor is used in turning a circular saw, for ordinary wood-cutting purposes. The exact loss of energy has not been ascertained, but Sir William Armstrong believes "there is every prospect of obtaining from 40 to 50 per cent. of the prime mover," or about $2\frac{1}{2}$ horse power on the saw in the above example.

In a similar manner, Dr. Siemens, F.R.S., utilises some dynamo-electric machines at his country seat near Tunbridge Wells. In this instance the primary power is derived from a steam engine, the waste steam of which is employed to warm the hothouses. During the night the primary machines are used in the production of two powerful electric lights, under the influence of which various fruits and plants are growing. During daylight the current from one of the machines is transmitted a distance of a quarter of a mile to the farm, when it is transformed into mechanical effect by means of an electro-motor, used to work the chaff-cutter and other machines. The current from the other machine is transmitted to the pumping-house, a distance of half a mile, and is then caused to pump the water required in the house, through an electro-motor, as in the former case. In this manner the fuel, which was formerly used in warming the hothouses, keeps the steam engine in motion, so that the actual extra cost for electric lighting and motive power is reduced to a minimum.

At their telegraph works, at Charlton, Messrs. Siemens employ the transmission of power by electricity for working the apparatus used in the mechanical testing of the cables. They also employ electro-motors for pumping, working cranes, hoists, and many other purposes.

These are but one or two examples of private enterprise in this direction carried out in England. In France the principle is extensively used, and dynamo-electric machines have for some time been employed in that country for working ploughing machines. But the application of the electrical transmission to railways has advanced much more rapidly than its introduction for driving stationary machinery. There are now on the continent of Europe alone a large number of independent examples of this class, particulars of which are offered in another chapter of the work.

As already hinted at, the application of the transmission of motive power will derive great advantages from the fact that, while large compound condensing steam engines may be made to develope the actual horse power with an expenditure of $2\frac{1}{4}$lb. of fuel per hour, very few small steam engines or locomotives will show a like result with an expenditure of 6lb. or 7lb. of coal. Thus, in considering the practicability of employing the electric current to distribute motive power from a central station to numerous surrounding points, it would be possible, even with an efficiency of 45 per cent. only in the distant motors, to produce power at those points much more economically than could be done by the use of separate steam engines. Danger, extra insurance rates, smoke, waste, and other inconveniences, not forgetting the wages of an attendant, would be done away with. The power could be obtained as easily as we now obtain gas, by the turning of a handle.

One central station could by these means supply all the motive power required within a radius of a mile, at a cost per horse power developed, at the most distant point, of less than 5lb. of coal. In London, and elsewhere, the electric lighting companies possess special cables for the supply of power.

The late demonstrations of the fact that electric energy can actually be *stored* or accumulated in suitable receivers have attracted a great deal of attention. In those instances it was shown that any quantity of electric energy might be accumulated in the secondary cells of Planté and Faure, and transported a thousand miles, to be reproduced as required. This is in itself a most important discovery, since not steam engines but also thermo-electric and zinc generators may be employed to yield the energy. It bears another significance also. It is possible to set a weak generator of current to store away its strength for, say, twenty-four hours in the accumulator, and to draw from the accumulator, when required, the twenty-four hours' energy in one hour, with twenty-four times the force of the primary generator. In this respect the prime source may be represented as slowly winding up a powerful spring, the power from which may be taken off as slowly as it was conferred, or very rapidly, and in great force, as required. The accumulators or condensers of MM. Faure and Planté present the additional advantage that they are most cheaply constructed, since only lead and lead oxide are employed, and that there is no loss of material in the process of charging and discharging. The accumulators may also be charged any number of times.

This valuable adjunct to the dynamo-electric machine has given a considerable impetus to both electric lighting and the transmission of motive power, for by its aid any required quantity or force of electricity may be kept in readiness, and supplied, as we have observed, under any conditions required. By its aid motive power may be conveyed or carried; but the accumulator is not likely to be used to any considerable extent in this way, save for purposes of supplying small powers.

Small moving powers may be produced by small currents, with considerable economy, more especially since late advances have opened a way for the storing of the energy of small powers. A very large number of minor applications of electricity as motive power have already sprung into existence, and there is every prospect that the number will be considerably augmented. The great advances which have been effected within the past few years have not only resulted in greatly-improved converters of the potentiality of zinc into electricity, but have also placed at the service of employers of small powers a description of motive engine of almost perfect design and construction. These improved converters of electric into motive energy are applicable in principle and practice as the most miniature motors, such as electric drills for the use of dentists, up to engraving machines, and motors for fans, ventilators, punkas, hair-brushes, or sewing and knitting machines. It is now practicable, at small expense, to actuate motors of this description, either directly by the currents from zinc cells or by means of accumulated currents in secondary cells, in turn derived from zinc or steam power.

It would appear that the electric light and the production of motive power should, to a great extent, go hand in hand. Many instances of the current for electric lighting being employed through the hours of daylight for motive power have of late sprung into existence. By these means, as some of the pioneers of electrical progress have observed, the electric plant is utilised to the utmost, and its value proportionately increased.

CHAPTER II.

On the Dynamical Nature of Electric Currents.

An impulse or discharge of electricity necessitates a transference or distribution of energy or work power. A current of electricity merely consists of a rapid and practically continuous recurrence of the simple impulse.

A current of electricity, therefore, implies a continuous transference of energy from one point to another. It also necessitates the expenditure or conversion of a portion of the energy—that is, at the beginning of the current, energy must be expended, and this is followed by a continuous absorption of a portion of the energy by the conductor, from the surface of which it is dissipated in the form of heat. Hence, the conductor *resists* the motion of electric energy, and a portion of the energy is necessarily expended in overcoming this resistance.

Unlike almost every other form of current from which work may be obtained, the electric current is usually utilised by taking advantage of its *inductive* effects. In the case of steam, the expansive force is applied directly to the piston of the engine. The same is done in the cases of water, gas, and gunpowder, in their respective fields of application; but in the case of electricity, as applied in the way of motive power, direct application of the current is useless. Motive power is obtained, as it were, indirectly, by causing the current to develope magnetic attraction in iron, the magnetic attractive force being applicable directly, as in the case of steam, to the moving parts of a machine.

A conductor, in the form of a wire, carrying a current of electricity, becomes heated by reason of the energy dissipated in overcoming its resistance. If the conductor were placed in a position favourable to the existence of inductive action, its temperature would be diminished in a degree proportionate to the amount of inductive action created. By coiling the current-bearing wire around a piece of iron, magnetic properties are developed in the iron, and the heat in the conductor is diminished in proportion to the work done by the magnetic attractive force.

Electro-motive Force.—These terms, as applied in text-books of electricity, refer entirely to the moving power of an electric impulse through a conductor. The tendency implied is also known as *tension*, and in the older treatises as *intensity*. It must not be confounded with the power of electricity as applied to the moving of machinery. It signifies, essentially, the tendency of electric energy to move from one point to another. In some respects it is analogous to water pressure, or "head" of water. A current of electricity is a transference of energy through a conductor under the influence of electro-motive force. Without potentiality or electro-motive force at the source of electricity there could be no conveyance or motion of electricity. A great quantity of electricity given off by the source may fail to flow through a conductor, unless the source also developes a certain electro-motive force. There are currents of low tension flowing from electric sources of feeble electro-motive force. Such currents pass but feebly through conductors offering sensible resistance. There are also currents of high tension flowing from an electric source of considerable driving power or electro-motive force. These flow vigorously and rapidly through highly-resisting conductors.

Resistance of the Conductor.—Electro-motive force is correlated with resistance, and is, under certain conditions, dependent for its existence upon resistance. There are no perfect conductors of electricity. Every metal used as a conductor offers a certain resistance to the passage through it of electricity. Resistance implies that property of the conductor by reason of which it prevents more than a certain amount of work being done in a given time by a given electro-motive force. It has been said that the amount of the passing energy expended on overcoming resistance is manifested as heat. Electrical resistance must not, however, be confounded with mechanical resistance, although heat appears in both cases. When water is forced through a pipe, the frictional resistance of the pipe is not constant. It varies with the quantity of water being forced through. In the case of electrical conductors, the resistance is *constant*, whether the passing energy be large or small in amount. In consequence of this, the calculation of electrical resistance is very easily carried out.

It is found that with a conductor of a given resistance conveying a current, an amount of electrical energy proportional to the electro-motive force passes. Upon doubling the electro-motive force, the strength of the current is doubled, or twice the amount of energy passes. From this it is clear that, with a constant resistance, the current is directly proportional to the electro-motive force of the electric source.

Conversely, keeping the electro-motive force constant, if the resistance be doubled, the current will be halved. Since the resistance depends upon the length of the conductor—the area being constant—the current will be inversely proportional to the length of the conductor. And since the resistance of conductors varies as the area, the same result may be attained by halving the area, which is exactly the same as doubling the length. In both cases the resistance is doubled. Again, the current is absorbed or retarded in proportion to the resistance of the conductor. The conductive power of the metallic path is inversely proportional to its resisting power. Hence, the more it resists the worse it conducts.

Electro-motive force, current, and resistance depend upon each other. The magnitude of one affects the amount of the others, and conversely. It is of much importance, in relation to the electro-motive machine, to fully understand this. The following formulæ may be employed in determining electro-motive forces, resistances, and currents. They are dependent upon the

c

principles previously enunciated, and are now generally known as Ohm's formulæ:—

If we use C for current, E for electro-motive force, and R for resistance, we find that C is proportional to E divided by R—

$$C = \frac{E}{R}$$

This, of course, assumes that the electro-motive force and resistance are known in relation to some unit of magnitude.

When the electro-motive force and the current are known, resistance may be determined by dividing the former by the latter—

$$R = \frac{E}{C}$$

Units of electro-motive force, resistance, and current have been determined by a committee appointed for that purpose by the British Association; but the absolute units so found are not used in practice. Multiples of them are, however, in common use. The unit of resistance is generally known as the ohm, and some idea of its actual magnitude may be obtained from the consideration that 485 metres of pure copper wire, one millimetre in diameter, at 0° Centigrade, present a resistance of one ohm. About 6ft. of pure copper wire of No. 36 Birmingham wire gauge also offers a resistance of one ohm.

The generally-recognised unit of electro-motive force is known as the volt. It is from five to ten per cent. less than the force manifested by the common Daniell telegraph-battery cell.

The unit of current or quantity of electricity or work-power actually flowing through a conductor is known as a weber.* An idea of its magnitude may be obtained from the consideration that *a weber per second* would flow in a circuit of *one ohm* resistance under an electro-motive force of *one volt*. Or, it may be assumed that a large Daniell cell would cause a weber per second to flow in a circuit of one ohm.

According to Dr. Joule, the *mechanical equivalent* of a weber current is ·735 foot-pounds per second, or 44·24 per minute.

Conductors and Insulators.—The enormous resistance offered by most non-metallic substances has caused them to be classed generally as non-conductors or insulators. All metals may be regarded as conductors of electricity, copper being universally employed for that purpose in the construction of dynamo-electric and electro-dynamic machines. Each body possesses a specific conductive power. For example, copper is a better conductor than iron, and silver a better conductor than copper. Silver is the best known conductor, but is too expensive to be generally used. Regarding the conductivity of silver as 100, copper may be taken as equal to 77·4, according to Wiedmann. Iron stands at 14·4, which places it out of the category of conductors for this class of apparatus.

Wires of copper used for electro-magnetic purposes are always insulated by being covered with some non-conducting substance, such as silk, cotton, hemp, or guttapercha, the latter forming the most efficient insulator.

---

* At the Congress of Electricians, held in Paris, in the autumn of 1881, it was generally agreed to term the weber an "Ampere."

Source of Electricity for Experimental Purposes. For purposes of experiment, or testing upon a small scale, the most convenient source of current consists of a voltaic cell or battery. The action of a single cell, of simple construction, will suffice to demonstrate practically the chief properties of the electric current.

When a pair of plates, zinc and copper, are placed separately in a vessel of acidulated water, no electrical manifestations occur. If, however, the upper edges of the plates be touched together, a current is immediately developed between and through the plates. This current is generally assumed to arise at the surface of the zinc plate, which, therefore, begins to dissolve. The current passes through the exciting liquid to the copper plate, by which it is conducted back to the zinc plate across the junction between them.

It is more convenient, and is generally indispensable, to fix a wire to the free extremity of each plate. It is also advisable to coat the zinc plate with mercury (amalgamation). The moving energy known as electrical current is assumed to arise, as before mentioned, at the zinc or dissolving plate. This movement takes place under the impelling agency of electro-motive force. The magnitude of the force, or, in other words, the tension of the cell, depends upon the ratio of chemical affinity between the acidulated liquid and the zinc and copper plates. The electro-motive force may thus be regarded as the result of the specific energy set free by the chemical combination of oxygen and zinc—a kind of furnace, in fact—less the counter electro-motive force of the copperplate.

The current, or quantity, of electricity a voltaic cell can yield depends upon the total resistance to be overcome. It will be equal to the quotient of the electro-motive force divided by the resistance.

Circuit.—It is absolutely necessary that the free extremities of the plates in the cell should be connected together, either directly or by means of a conductor, before any manifestation of electricity is yielded. This necessitates the moving of the energy in a *circuit*. If there should occur a break of conductive continuity, however minute, at any part of the circuit, all electrical phenomena will cease, and the action of the generating cell will be suspended. We must, in fact, provide a conducting circuit.

Electric Poles.—These terms are applied to the extremities of the voltaic plates, or to the extremities or wires of any source of electricity. In the case of a voltaic cell, the wire attached to the copper plate is known as the *positive* pole, since it conducts the current *from* the combination. The wire attached to the zinc plate is known as the *negative* pole, since it receives or conducts back the current. The same terms are applied, whether the poles be those of a single cell, a battery of cells, or a dynamo-electric machine. The positive pole is always that from which the current is assumed to issue. This determines the direction in which the current moves in the circuit. The negative pole is always that by which the current is assumed to return to the generator. The sign $+$ is conventionally used to indicate the positive pole, and the sign $-$ to indicate the negative pole.

In the voltaic cell itself the plates are termed positive and negative in relation to each other. The active plate (zinc) is known as positive, the passive plate (copper) being negative.

The term poles is also used in speaking of the extremities of a piece of magnetised steel or iron. The terms positive and negative are not, however, generally employed to indicate particular extremities of magnets. North pole and south pole are much more common, with reference to the geographical position assumed by a bar magnet when free to rotate.

Ascertaining the Direction of an Electrical Current. When the poles of an electric source are connected together by a wire, a current is set up in a particular direction through the wire. It is assumed to flow from the electrically-positive to the electrically-negative pole. There is no easy and generally-applicable means of distinguishing the poles, save by means of a magnetic needle, which deflects in a particular direction, previously determined, when a current passes parallel to it. If the conducting wire be cut, and the two poles dipped for a moment in a solution of sulphate of copper, the positive pole will commence to dissolve, and the negative extremity will *receive a coating of copper*. This of itself, a most simple experiment, is sufficient evidence that the current is moving in a particular direction, and the nature of the result indicates almost conclusively that the current flows from the positive to the negative pole.

Illustrations of the Dynamical Effects of Electric Currents.—When the poles of an electric source are connected together by means of a long, insulated wire, a resisting medium for the flow of the current has been provided. The energy of the current will in this case be chiefly expended in heating the total resistance in circuit—that is, the whole circuit, through generator and interpolar resistance, will be raised in temperature. But it will be particularly observed that the interpolar is not the only resistance. A certain resistance is offered by the portion of the circuit through the *generator* itself. This is generally known as the *internal* resistance of the source. In the case of a voltaic battery, it is due chiefly to the indifferently-conductive qualities of the exciting liquid. In machines it is chiefly due to the wire coils. When no work is being done by the current the whole energy produced is expended in heating the circuit.

As an example of the effects above mentioned, it may be observed that a powerful voltaic battery or dynamo-electric machine, when allowed to work upon short circuit, frequently makes the interpolar wire red hot, and burns the insulating material. In cases where the current is strong enough the conductor is melted. When, however, work, of any of the kinds about to be spoken of, is put into the circuit, the temperature falls, and the energy is diverted. Hence, it is comparatively easy to obtain proof of the general statement, that *the energy of an electrical current is convertible into heat.* In fact, the mechanical equivalent of the available electric energy in circuit may be determined by means of the heat produced. If the interpolar wire be coiled up and immersed in one pound of water, the temperature of the latter will be raised, and each degree (F.) of heat so obtained may be regarded as equal to 772 foot-pounds of energy. These figures are derived from the determinations of Dr. Joule. Experiment shows the heat developed to be proportional to the squares of the current strengths, and also to the resistance offered by the wire.

Induction.—The greater portion of the following examples of the dynamical and inductive effects of the current may be produced by the aid of a small generator of the voltaic type, or the simple apparatus already mentioned.

Induction is a term applied to a property of a current which tends to give rise to other currents or to magnetism. Under certain conditions, a current tends to set up an opposing or reverse current in a neighbouring closed circuit. In doing so, the primary current parts with a portion of its energy. In the case of two circuits laid side by side, when a current commences to flow in one, it will *induce* an *inverse* current in the other. But this induced current is only momentary. It ceases as soon as the primary current is fully established. It does not flow continuously, like the primary current. If the primary current be stopped, a *direct* induced current in the secondary wire will mark the change. Any change in the condition of the primary current will affect the secondary, by setting up a current in it. If the primary current wire be approached, an inverse current will flow in the secondary; if it be withdrawn, a direct current will be produced. All of these currents are but momentary, or they exist only as long as the primary wire or current is undergoing some change. No induced current is produced by a continuously-flowing primary current in a wire at rest, when the secondary wire is at rest also.

These peculiar effects are most conveniently observed by coiling up the primary wire, and causing it to pass into a helix, made by coiling up the secondary, the extremities of which should be attached to a galvanometer, to enable the induced currents to manifest themselves.

Energy is, of course, given up by the primary current at starting and cessation. Whether this loss of energy is proportional to the amount appearing in the secondary wire depends upon the nearness of the conductors and upon other conditions.

Production of Mechanical Motion.—When a magnetised needle, free to rotate, is brought near and parallel to a current-bearing wire, it is deflected to a position tending across, or at right angles to, the wire. The same effect takes place if the needle and wire be placed in the magnetic meridian, N. and S., and the current then passed. The extent of deflection from the magnetic meridian depends upon the strength of the current. The ratio of deflection to the strength of the current is nearly as the tangents of the angles produced.

Upon this important principle *galvanometers*, for ascertaining the strengths of currents, are constructed. When the circuit wire is coiled several times around the needle, so as to yet allow the latter freedom to rotate, the directive force of the current is increased nearly as the number of turns.

Magnetic Induction.—Electro-Magnetism.—If the current-bearing wire be coiled in a helix around a piece of soft iron, magnetism will immediately be produced in the iron. It will become a magnet, and will act exactly as a natural or permanent magnet. Its magnetism is not, however, permanent. It vanishes upon the withdrawal of the current. It can be instantly recalled by restoring the current. In this manner the bar may be magnetised and demagnetised as often as required, and with extreme rapidity.

This property appears to belong almost exclusively to *soft* iron, for when cast iron, or even soft steel, is used, a great portion of the induced magnetism is retained for some time, and some of it permanently. In this manner, when hard steel is magnetised by the current, about 90 per cent. of the induced magnetism is permanently retained. It is noteworthy, however, that, as the iron or steel increases in hardness, less and less magnetism is produced by a given current. A magnetic force of 100 in pure Swedish iron could not by an equal current (the current strength being moderate) be imparted to cast iron or steel. For cast iron the magnetism induced might manifest a force of 70, and for hard cast steel probably not more than 50.

It is provided, in the above statement, that the current strength is assumed to be moderate, a provision which calls for further explanation. There is a limit to the intensity of the magnetic polarity which may be imparted to iron and steel. When an iron bar is subjected to the inductive influence of a powerful current, it becomes what is known as *saturated*, and cannot be magnetised to a greater extent, whether the current be made stronger or not. The saturated stage varies greatly with the material. No increase in the power of the current can be made to carry the magnetising process further. Hence, as above stated, if an abnormally strong current were used, both iron and steel would become saturated, but the steel would require the larger expenditure of energy to accomplish this.

A certain expenditure of *time* is also essential in the process of magnetisation. The time also varies with the material. Very soft iron may be fully magnetised in so short a time that the change may be regarded as instantaneous. In the case of hard iron and steel, a sensible time must elapse before the fully magnetised condition is attained. The time required also varies with the strength of the current employed. A powerful current will magnetise even hard steel in a fraction of a second; but the usual current strength employed for the purpose does not induce the fully magnetised condition in hard steel in less than one second.

The time necessary to effect complete self-demagnetisation does not appear to bear any known ratio to the time expended in magnetisation. Very soft iron can be magnetised instantaneously, and the self-demagnetisation is also practically instantaneous. Hard iron and steel may be magnetised almost as rapidly, but the process of demagnetisation is very slow in the case of hard iron; and for hard steel demagnetisation may never take place.

When soft iron is magnetised by the current, a portion of the energy is expended in the process. This energy is stored upon the molecules of the iron, and when the current is suspended it is returned in the form of an inverse induced current in the conducting helix. In the case of hard steel, the energy so stored upon the molecules is retained, and is known as permanent magnetism. Hence the inverse induced current is but feeble when hard steel is treated.

It should be particularly noted that a permanent or electro-magnet *possesses practically no energy of its own*. It consists of a mass of steel or iron in which the molecules are assumed to be arranged in a peculiar manner, known as polarity. It is not, as is frequently erroneously assumed, a magazine of force. It can only give up at most what has been imparted to it, and in attracting an armature it exhausts its energy. Its power is restored by forcibly withdrawing the armature.

When a bar of soft iron is rapidly magnetised and demagnetised, under the influence of an intermittent current of considerable strength, a great portion of the energy so expended appears in the bar and exciting coil as *heat*. This is more easily observed if the bar be subjected to reversing currents rapidly following each other. Rapid reversals of magnetic polarity therefore give rise to heat. A practical example of this property may be observed in some forms of dynamo-electric machine, where the revolving armature is subjected to rapid reversals of polarity. In many cases the armatures have become so hot as to burn away the insulating covering of the wires. Some of these machines are artificially cooled to dissipate the heat generated in working. In other forms of machine, when a great many reversals take place during each revolution, as in the Siemens reversing machine, the iron cores are entirely dispensed with, induction in the wire coils only being depended upon.

The property of soft iron to rapidly absorb and release charges of magnetic energy is largely taken advantage of in every useful form of

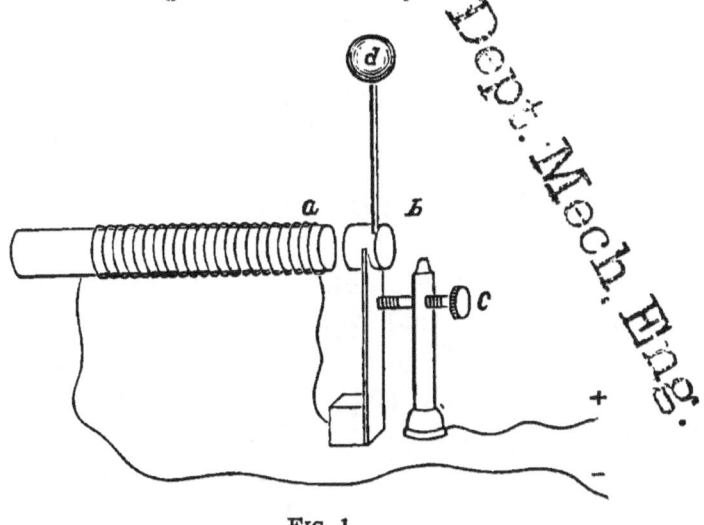

Fig. 1.

dynamo-electric and electro-dynamic machine. This principle is frequently taken advantage of in scientific and telegraphic apparatus, in the construction of a most simple form of electro-motor, which may be devised as follows: A bar of soft iron a (Fig. 1) is mounted upon a stand. Around the bar are coiled two or three layers of insulated copper wire. Opposite one end of the bar is also mounted an iron ball or armature b, attached to an upright spring. Behind the armature is an adjustable contact or regulating screw c, working through a metallic pillar. One extremity of the electro-magnetic coil is attached to the armature spring, the other leads to one pole of the exciting cell supplying the current. The reverse pole of the cell is attached to the screw pillar, and the arrangement is complete.

The action of this arrangement is simple and instructive. As soon as the current passes, the electro-magnetic bar is excited, and attracts the armature towards it. But this movement, by pulling the spring away from the contact-screw, breaks the circuit, and the bar becomes demagnetised by the cessation of the current. It therefore releases the armature, which falls back to its former position, again completing the circuit. By these means the armature is kept vibrating with great rapidity.

By means of this device, motion might be obtained by attaching a rod and crank to the pendulum spring or armature. If the electro-magnetic bar were curved to a U form, so as to bring both poles to bear upon the armature, greater force would be obtained. This device is employed for interrupting the circuit in induction and other apparatus. The armature may also be used as a bell-hammer d, striking the bell at each vibration. The arrangement is used extensively in electric bells.

Since it is found that an increase in the number of turns taken by the current-bearing wire around a magnetic needle increases the deflective force almost in proportion, it is also clear that by increasing the turns of the helix around electro-magnetic bars the magnetic force induced will be increased. Hence, when the current is of moderate strength, and the iron bar of moderate dimensions, the magnetism conferred upon its molecules is *in proportion to the strength of the current and the number of turns in the exciting coil.* It is also found that, under favourable conditions, the weight which the magnet sustains is in proportion to the squares of the current strength. This rule assumes that the saturated condition has not been attained.

The forms which may be given to electro-magnets are exceedingly numerous. Tubes, bundles of wires, bars, or plates may be used for cores. The enveloping wire may be fine or thick, according to the electro-motive force and internal resistance of the electric source. This portion of the subject will be found treated at greater length in the succeeding chapter.

Production of Magneto-electric Currents.—The production of electric currents by the inductive influence of magnets is so closely allied to the production of magnetism by currents that it deserves the closest attention.

When a magnetised steel bar is quickly passed within a wire helix or solenoid, forming a closed circuit, a rapid rush of current denotes the inductive effect which takes place. The mere fact that a current is induced may be rendered apparent by placing in the circuit a sensitive galvanometer. If the magnet be withdrawn from the solenoid, a similar, but inverse, current will indicate the movement. These currents are only momentary.

They exist only while the magnet is in motion relatively to the solenoid. Work or energy is expended in both instances. This expended energy is transformed into an electric current. Therefore, it is less difficult to move the magnet in the coil when the circuit is incomplete than when it is complete. Since the current does no external work, it may be assumed that in this case it is expended on the resistance of the conductor, and appears as heat.

As may be inferred, the number of turns taken by the conductor in forming the solenoid has a great influence upon the electro-motive force of the induced current. The electro-motive force increases up to a certain point almost directly as the number of turns in the coil.

In the construction of apparatus for generating magneto-electric currents, the arrangement described above is seldom employed. If a permanent magnet be brought near to a piece of soft iron, the latter instantly becomes a magnet also by induction. The magnetic polarity will be the reverse of that of the steel magnet. Of course, the same takes place if an electro-magnet be used in place of a permanent magnet.

Since a permanent or electro-magnet can induce magnetism in soft iron, it follows, from the electro-magnetic laws already enunciated, that upon withdrawing the inducing magnet all effect upon the soft iron ceases. If the soft iron were placed within a helix, with closed circuit, it would be found that an electric impulse in the wire would mark each magnetisation and demagnetisation. These impulses would be reversed to each other, and they would only exist during the time occupied in magnetising or demagnetising. In this respect they would resemble induction currents created by other means.

It is of little or no consequence by what means the soft iron magnet or armature is moved in relation to the inducing magnet. It may be made to approach towards and withdraw from it in a straight line, or the poles of one may be rotated near to the poles of the other. A series of electric impulses would thus be caused to flow in opposite directions in the induced circuit.

Upon this principle the action of several ingenious magneto-electric machines is based. They all consist of devices for taking advantage of the inductive property of magnetism in soft iron armatures upon which wire is coiled. Following the law that the electro-motive force developed is to a certain extent dependent upon the number of convolutions described by the conductor around the armature, it follows that, in this class of apparatus, the longer the wire the greater the electro-motive force developed.

It should be clearly understood, however, that the inductive influence of a magnetised armature is only exercised within narrow limits of space, so that, as the number of turns of wire required becomes greater, finer wire must be used, otherwise power will be lost by the resistance of convolutions outside of the influence of the magnetised core. The current or quantity becomes increasingly smaller as the wire becomes finer, and as the length is increased, but the electro-motive force increases.

D

Machines based upon these principles have within the last fourteen years been brought to a wonderful degree of perfection, the currents developed by the larger machines, driven by steam, being so powerful as to yield numerous centres of electric light, and to fuse up considerable lengths of thick steel bar.

It will be particularly observed that electro-magnetic and magneto-electric phenomena are almost perfectly the converse of each other. Most of the magnetic effects produced in soft iron by an electric current have, conversely, a striking analogy in the production of currents by magnets. In the first case, an increased length of wire in the coil, while it augments the influence of the current, calls, by reason of its increased resistance, for higher electro-motive force to maintain the current. In the second case, the electro-motive force increases with the length of wire—one effect being thus the converse of the other.

Magnetic Tension and Electro-motive Force.—In the case of moving and inducing a permanent magnet near to a soft iron armature or inductor, as exemplified above, the electro-motive force developed in the wire coil will depend not only upon the number of turns described by the wire, but upon the intensity of the magnetic field set up, and upon the velocity of movement. With a given number of turns in the coil, and a given velocity of the armature, the electro-motive force developed may be varied by varying the intensity of the magnetic field through which the armature moves. Hence, it is of great importance, since the magnetic intensity decreases as the square of the distance increases, to cause the opposed polar faces to pass as near to each other as possible without actual contact. Again, with a given number of turns in the wire coil, and a given magnetic field, the electro-motive force may be varied by varying the velocity of rotation. Experiment does not appear to show that there exists any constant ratio between the velocity and the electro-motive force developed. The force increases, as a rule, more rapidly than the velocity, up to a certain point, dependent upon the magnetic field and the purity and softness of the iron.

Internal Resistance of Magneto-electric Inductors. The resistance of the wire coiled upon the soft iron armature may be regarded as internal resistance, in the light of the internal resistance of a voltaic cell, as previously spoken of. It is essential that the machine should have a coil of many turns, so as to develope considerable electro-motive force (this involves considerable internal resistance), to yield a sensible current through an interpolar resistance of any considerable magnitude.

CHAPTER III.

Electrical Measurements.

IN the preceding chapter the various correlated laws and properties of currents and magnetism were spoken of. It is now proposed to show how electrical magnitudes—as electro-motive force, current, and resistance—may be dealt with in practice, so as to confer upon the terms *current* and *resistance* some definite meaning, and to bring them, as far as practicable, under a system of measurement. It may, indeed, be generally assumed that a system of measurement dealing with electrical magnitudes is almost as indispensable to the student of electro-dynamics as a system of measurements by mechanical dimensions can be to the mechanical engineer. The practical man must generally know how much current, how much resistance, or he knows nothing.

There are several difficulties in the way of a generally applicable system of measurement adapted for powerful currents, such as are yielded by dynamo-electric machines. Smaller currents, as those from voltaic batteries, are comparatively easily dealt with. In the measurment of resistance the difficulties are also comparatively easy to overcome. As an aid in this branch, a table, giving a list of electrical wires of various gauges, with their respective resistances in feet per ohm is given at p. 61.

Instruments Used in Measurement.—The most important instrument in general use for measuring electrical magnitudes is the galvanometer, the principle of which has already been explained. The tangent galvanometer is almost the only form of the many varieties in use which may be employed with advantage in measuring powerful currents.

Common galvanometers, with long needles, cannot be depended upon. They serve very well, however, as detectors of currents, and for comparisons in a certain degree. When the needle is very short, and the exciting coil very large, the deflections have a definite value in relation to each other. The ordinary form of the tangent galvanometer consists of a hoop of copper, frequently twelve inches in diameter, so arranged that the current may be passed around it. The needle is free to rotate on a pivot in the centre of the hoop. It may be assumed that, when thus arranged, the motions of the needle will not alter its position relatively to the disturbing power of the ring. When thus arranged *the tangents of the angles to which the needle is deflected by the currents are proportional to the currents causing the deflections.*

In this manner the tangent galvanometer may be employed to measure a current under different conditions, through varying resistances, and with varying electro-motive force. It may likewise be used to compare currents together, or to compare currents with some recognised standard of deflection. A good deal of practical work may be done in this way without reference to units. When it is required to measure currents in terms of the electrical unit, it is advisable to ascertain the deflection value of a unit current, and to indicate the point as a fiducial mark upon the graduated dial.

Obach's Tangent Galvanometer. — Among the many new electrical instruments exhibited by Messrs. Siemens at the Paris Electrical Exhibition of 1881 was Obach's tangent galvanometer, specially devised and constructed for the measurement of very powerful currents, as used on electric railways, electric light and other circuits.

In some respects the instrument is similar to the common tangent galvanometer. The ring through which the current passes is movable around its horizontal diameter lying in magnetic meridian. The inclination of the ring to the horizontal plane is read off on a vertical scale. With a constant strength of current, the force with which the magnetic needle in the centre of the ring is deflected from the meridian is proportional to the *sine* of the angle which the plane of the ring encloses with the horizontal plane. Instead of measuring the angle between the ring and the horizontal plane, the angle which the ring makes with the vertical plane could also be taken, but in this case the *cosine* of the observed angle would be employed in place of the sine. In this instrument the angle of the ring with the horizontal plane is read off, as the natural values of the sine are more frequently given in tables than those of the cosine, and can be used without further trouble. In the engraving (Fig. 2) a perspective view is given, showing the ring in an inclined position.

From a base G, provided with three levelling screws, rises a stout brass column S. The bottom of this column is formed of a square block, cut out at one side in the shape of a ⊃, and made strong enough to insure it against bending. The column can be turned on its axis in order to place the instrument in the magnetic meridian. The screw S^1 holds it firmly in its place. On the top of the column is a circular brass box N, about 8in. in diameter and

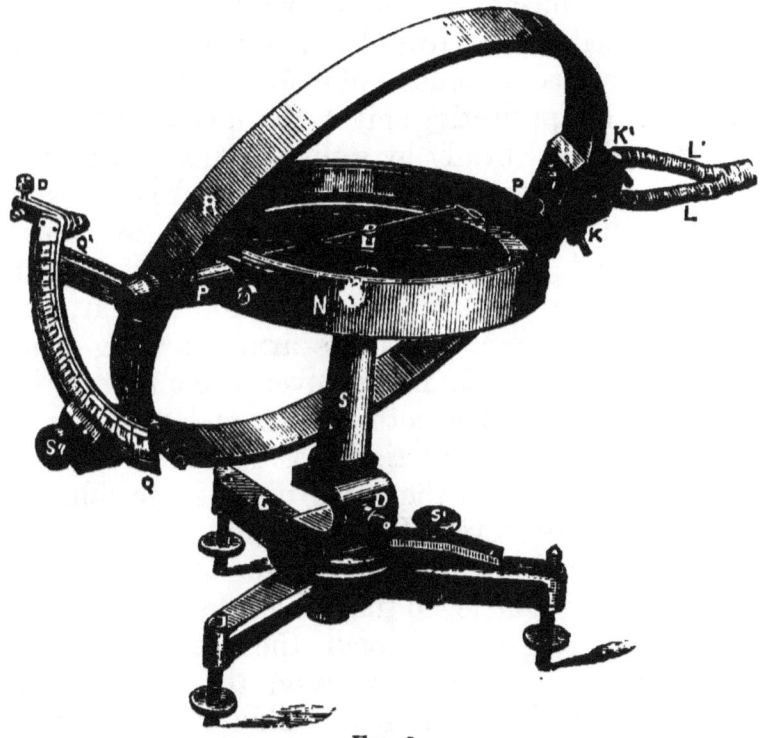

Fig. 2.

1¾in. high. Conical arms P P^1 are screwed to opposite sides of the box, and serve to support the ring R. This ring is made of gun-metal, containing a high percentage of copper. The dimensions are 11¾in. inner diameter, about ⅜in. thick, and 1in. broad, so that the ring offers an exceedingly small resistance to the current. At P the ring is cut through. On the outer face and at each side of the opening a semicircular piece of gun-metal, with clamp screws K K^1, is fixed. On the inner face of the ring the opening is bridged over by a strong piece of ebonite H, in which a brass bush is inserted to receive the pivot on the arm P. The pivot on the arm P projects somewhat beyond the ring R, and is insulated therefrom by an ebonite collar, over which is placed a brass collar for the ring to turn on. On this projecting pivot the quadrant Q^{1}Q is placed, and held firmly by a pin. This quadrant is divided into degrees, and the angles, the sines of which (for radius=1) correspond to the values 0·1, 0·2, 0·3 . . . 1·0, are shown on the same scale by longer marks. Fixed to the ring R, and insulated therefrom, is a brass arm, carrying at its end a vernier, which moves over the quadrant Q. The screw S^{11} clamps both vernier and ring. The latter at its extreme vertical position is arrested by a screw D, shown in the lower part of the column, and by another screw D^1, when in the horizontal position. On loosening the screws Q the vernier can be shifted a little, so that when the zero adjustment of the ring is first made the required agreement between the divisions on the vernier and quadrant can be obtained. This adjustment is made by sending as strong a current as possible in alternate directions through the ring and then bringing the latter by alteration of the screw D^1 into such a position that no deflection of the needle takes place. When this is the case, the vernier is set to zero. Within the circular box N, and at about half its height, is the dial circle divided into degrees of sufficient size to enable a tenth to be easily estimated. The length of the magnet (representing the magnetic needle) is only one-fifteenth the diameter of the ring. It carries an aluminium pointer, and is fixed to an axis, which is worked to a fine point at each end. On the circle dial, in an exact line through the bearings of the ring, is screwed a light brass frame, having an agate bearing in the centre, and beneath it, in the middle of the box, is a second jewel or agate bearing. In these bearings the axis of the magnet plays. A screw underneath the brass box, acting on a brass spring, stops the movement of the needle. A small spirit-level is fixed in the box, so that the galvanometer may be accurately levelled. The current to be measured is led to the ring by means of two thick insulated copper wires $L^1 L$, which are wound together for a distance of about three feet, in order to obviate the possibility of the current in the wires themselves deflecting the needle. The extremities of the wires are connected to the clamp screws $K K^1$. If greater accuracy is required it is advisable to take readings on both sides of the zero position, viz., to change the direction of the current in the instrument by means of a suitable commutator.

The following three cases illustrate the various methods of measuring currents with this galvanometer:—

1. Currents of *different* strength, $I_1, I_2, I_3, \ldots$ sent through the ring at the *same* inclination ϕ, showing deflections a_1, a_2, a_3, give—
$$I_1 : I_2 : I_3, \ldots = \tan. a_1 : \tan. a_2 : \tan. a_3 \ldots$$
viz., the law of the tangent holds also for the inclined ring.

2. The *same* current I sent through the ring at *different* angles of inclination $\phi_1, \phi_2, \phi_3 \ldots$ gives—
$$\tan. a_1 : \tan. a_2 : \tan. a_3 \ldots = \sin. \phi_1 : \sin. \phi_2 : \sin. \phi_3 \ldots$$
or $\dfrac{\tan. a_1}{\sin. \phi_1} = \dfrac{\tan. a_2}{\sin. \phi_2} = \dfrac{\tan. a_3}{\sin. \phi_3} = \ldots = $ constant.

42   ELECTRO-MOTORS.

The tangents of the deflections are therefore in the same proportion as the sines of the inclinations; or, in other words, the tangents of the deflections divided by the sines of the corresponding inclinations give for the same strength of current a constant value.

3. For different currents $I_1 : I_2 : I_3$ . . . sent through the ring at inclinations ϕ_1, ϕ_2, ϕ_3 . . . giving the same deflection a (say of 45 deg.) we have—

$$I_1 : I_2 : I_3 \ldots = \frac{1}{\sin. \phi_1} : \frac{1}{\sin. \phi_2} : \frac{1}{\sin. \phi_3} \ldots$$
$$= \text{cosec.} \ \phi_1 : \text{cosec.} \ \phi_2 : \text{cosec.} \ \phi_3 \ldots$$

and the instrument thus used acts as a *cosecant* galvanometer.

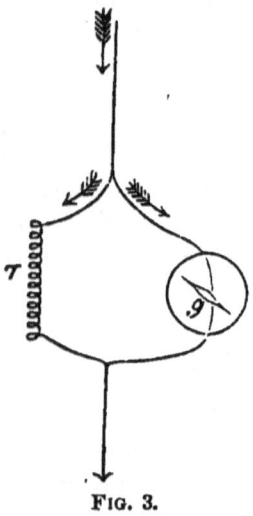

FIG. 3.

Shunts are small resistance coils supplied with galvanometers for the purpose of varying their sensibility. The shunts usually have resistances of $\frac{1}{9}$th, $\frac{1}{99}$th, and $\frac{1}{999}$th of the galvanometer coil itself. Therefore $\frac{1}{10}$th, $\frac{1}{100}$th, or $\frac{1}{1000}$th of the whole current may be sent through the galvanometer (Fig. 3). In the measurement of large currents, as those from dynamo-electric machines, a shunt of such resistance as will bring the deflection to 45 degrees is usually employed. Although there is considerable liability to error in the use of shunts, yet the simplicity of the tangent galvanometer has much to recommend it when compared with the electrometers, potentiometers, electro-dynamometers, frequently used for measuring powerful currents. Moreover, the construction and use of these instruments do not come within the scope of the present treatise. The instruments are made by Messrs. Siemens, of Westminster, and other electricians.

Resistance Coils and the Rheostat are used for comparing, and so measuring resistance. Resistance coils are usually of fine German-silver wire in lengths equal to known and stated numbers of ohms. They are generally arranged in a case, as sets, with connections, so that any required resistance may be thrown into or withdrawn from a circuit. The *rheostat* is also a resistance instrument, chiefly useful for introducing and withdrawing small resistances. These instruments are generally used in resistance measurement, in conjunction with a galvanometer and a current from a cell or two of a voltaic battery.

Measurements of Resistances.—Cases frequently arise when the ordinary calculations of resistances, reckoning by length and size of wire, are entirely inapplicable. In such instances it becomes necessary, when a resistance should be known, to employ some method of measurement. Calculation is also inferior to measurement in point of accuracy, since the conducting metals used vary in different samples, often to the extent of 20 to 50 per cent.

Perhaps the most useful source of the small current necessary in determining resistances is the Daniell cell previously mentioned. It consists of a cylinder of copper, plunged in a suitable vessel containing a solution of sulphate of copper. Within the copper cylinder is placed a narrow unglazed or porous cell, containing a rod of zinc in acidulated water. The zinc rod should be coated with mercury, to prevent local action, and wires should lead from both metals for purposes of connection. The electro-motive force of this cell is 1·079 volt, in terms of the unit. If the internal resistance of this cell were small, it would be capable of yielding a current of one weber, but the internal resistance is usually as great as two or three ohms. A cell of this kind, constructed so as to fill a jar capable of containing a gallon of liquid, would probably yield in a small resistance a weber current.

The internal resistance of the voltaic cell once known, greatly simplifies the process of measuring a resistance by the galvanometer method. In order to ascertain the internal resistance of the Daniell cell, produce a deflection on the tangent galvanometer. (For small currents, tangent galvanometers are generally provided with a long thin wire coil.) Insert in the circuit such known resistance as will bring the needle to a convenient angle. Note this deflection, and add further resistance until the angle of deflection is reduced in value to exactly one-half, calculating by the tangents. This doubles the resistance, and the added resistance (the latter addition) is equal to that of the cell, less, of course, the galvanometer and the first added resistance. The galvanometer resistance should be either equal to nothing, or, if measurable, known, and subtracted from the result.

In like manner, the total resistance of any circuit may be determined by inserting in it a voltaic cell of known internal resistance. By producing a deflection in the circuit to be measured, noting it, and adding known resistance in ohms, until the current indicated by the galvanometer is exactly halved, it is clear that the number of added ohms is equal to the resistance previously in the circuit, less the cell. The simplicity of this method is based upon the law that the current strength is inversely proportional to the resistance.

By the aid of the wire table, given at p. 61, the resistances of known lengths and sizes of copper wire used in common circuits may be readily calculated. From 10 to 20 per cent. is frequently added to the figures given (for pure copper), in order to calculate the resistance due to ordinary commercial wire. In such calculations it will prove of service to bear in mind the rule that *the resistance of a wire of constant section and material is directly proportional to the length and inversely proportional to the area of its cross section.* The conductor may be of any shape. It is a matter of indifference what form the cross section exhibits, because the resistance is in no way affected by the extent of the surface of the conductor. Hence, conductors may be round, square, flat, tubular, or of any other required section.

Several other methods of measuring resistances are in common use. The Wheatstone bridge method is one of the most trustworthy, but a description of it does not come within the scope of the present work.

Determination of Electro-motive Force.—The electro-motive force due to any source of electricity can be determined by calculation, when the resistance and current are known, by the method previously spoken of—

$$C \times R = E$$

Under unit measurements (webers and ohms) this gives the electro-motive force in volts. It is often necessary, however, to determine the electro-motive force in cases when neither the current or the resistance of the circuit is known. It is usually done in terms of some known standard of electro-motive force. The Daniell cell is the most convenient standard for ordinary use. Its force is as nearly as possible 1·079 volt, when acidulated water is used in the zinc compartment. Ordinarily, the force of this cell may be regarded as 1 volt, the small fraction ·079 being neglected.

By means of a tangent galvanometer, provided with a long coil of fine wire (the resistance of which need not be known), the force due to any source of electricity giving currents is proportional to the tangents of the angles of deflection. Having determined the angle described by source of known force, as the Daniell cell, comparisons of smaller or larger currents may be made with it.

Thomson's electrometer is now generally used in determinations of electro-motive force, and Clark's potentiometer is also extensively employed, but the calculations necessary involve so much special knowledge of electrical potential that they cannot be treated here.

The strength of the current due to any electric source may be determined by the heat developed in the circuit. This method has been used by Dr. Siemens, Dr. Hopkinson, and others. The heat developed by a current C, in a resistance r, in time t, is C^2rt. Thus the current may be measured by the quantity of heat received in a given time by water in which a resistance r is immersed. A wire of known resistance is immersed in a given quantity of water, placed in a calorimeter, and the temperature of the water is measured. Dr. Joule found that $H=C^2rt$, where H is the quantity of heat produced by the current.

CHAPTER IV.

Electro-Magnetic Force.

NUMEROUS experiments, conducted by various authorities, have failed to show that the electro-magnet, in reference to the force it yields, is governed by laws permitting the direct use of accurate rules and formulæ.

All the formulæ applicable to the laws under which electro-magnetism is developed relate to three elements :—

    1. The iron core.
    2. The encircling coil of insulated wire.
    3. The current passed.

Great diversity exists between the results published by different authorities, differences which long experience would appear to attribute to the kind and quality of the iron used, to its form, and to other conditions which it is impossible to provide for in general rules. The following formula is useful :—

    Let M = the magnetic force of the electro-magnet.
        n = the number of convolutions (wire).
        d = diameter of the iron core.
        C = the current passing.
        c = a constant (not stated).
        Then $M = cnC\sqrt{d}$.

Jacobi found that "the magnetism of an electro-magnet is directly proportional to the number of turns of the helix." To this should be added, the currents being equal.

Dub found that " its attraction is proportional to the square of the number of convolutions," and that " the attraction between two electro-magnets is proportional to the sum of the product of the current strength and number of convolutions of both helixes."

Menzzer found what is well known according to another law, that " when the resistance of the coils of the electro-magnet is equal to the resistance of the rest of the circuit, the magnetising force is at a maximum."

The truth is, that there is little difficulty in constructing electro-magnets to yield as mechanical effect about 90 per cent. of the energy of the current. The magnetising power of a coil will be at a maximum when its resistance is equal to that of the electric source, no matter of what size and length of wire it is composed. The softness of the iron core only affects the result as follows : a soft iron core is magnetised more rapidly and fully than a hard core, and a small core is magnetised more rapidly than a large one.

The size of the wire used in the magnetising coil must obviously depend upon the internal resistance of the electric source, and the electro-motive force. The number of turns it describes around the core will thus depend both upon its size and upon the resistance at the source, since it is not found advantageous to increase the number of layers of wire when a thickness equal to the diameter of the core has been attained.

In developing electro-magnetism in any mass of

E

iron, it must be observed that the effect of each successive layer of wire at increasingly greater distances from the iron rapidly diminishes. Therefore it is advantageous to employ a wire small enough to give a considerable number of turns around the core in a very limited radius. The magnetism developed by an electro-magnet thus arranged will be proportional to the current strength and the number of turns in the coil. When the core is small, say less than one third of an inch in diameter, a maximum is soon reached beyond which additional turns of wire yield no additional magnetism; and with a given number of turns of wire a maximum is soon attained beyond which it is not advantageous to increase the current strength—this is the maximum of *magnetic saturation*.

The length of the electro-magnetic bar only serves to insulate its poles, but this condition greatly influences the nature of the magnetic field. It is found that it is of little moment in what way the wire is wound, so long as it is maintained at right angles to the axis of the bar, whether it is spread from end to end of the bar or accumulated at the extremities. It must not be overlooked, however, that the *magnetic poles* will tend to lie near to the points of greatest excitation. Hence, it is more advantageous to accumulate the wire, under the above rules, near to the extremities of the bar.

Magnetic Attraction will depend upon the magnetic field set up, and this varies in extent, to a certain degree, with the distance apart of the poles. A horseshoe electro-magnet, having a space of two inches between the centres of its poles, will carry a greater weight attached to its armature than it would were its poles separated to four inches. But in the first and second cases the distance from which armature might be attached, or, in other words, the extent of the magnetic field, would not be the same. The extent of the magnetic field may be shown (by means of iron filings sprinkled upon paper) to occupy an ellipse, and the extreme radius of this ellipse can be shown to be at a greater distance from the poles as they are wider apart. The difference, however, is not very great, but it frequently upsets the general conclusion that, the magnetic attraction is inversely proportional to the square of the distances between the magnet and its armature.

Armatures should be considered in the light of induced magnates. It is therefore important that, where the magnet is intended to act upon the armature in the axial line, the armature be large enough in section to engross the inductive force in the field, or at least equal in section to the core itself.

From the most elementary conceptions of magnetism, it is obvious that when a magnet induces a magnetic charge upon the molecules of its armature the polarity is of reverse name, otherwise repulsion, and not attraction, would ensue. The end of the armature opposite to the N pole of the magnet will become of S polarity, and the opposite extremity of N polarity.

According to the same law, when, as is frequently the case, the armature is an independent magnet by itself, the poles must be arranged, when it is required that attraction should take place, according to opposed polarities, N opposite to S, and N^1 opposite to S^1. Further, it is clear that in the case of also utilising repulsion, as is sometimes done, it is only necessary to change the magnetic polarity of one of the pairs, so that S shall face S^1 and N N^1. When the current is reversed in the magnet the polarity is reversed also.

Residual Magnetism is to be found in all masses of iron once magnetised. Its amount will depend upon the hardness of the iron and upon the time during which the armature has remained in metallic contact with the poles. It is found that the latter condition, forming as it does a closed magnetic circuit, is strongly favourable to the development of a permanent magnetic charge upon the molecules of the iron. In large masses of iron the residual magnetic influence is often very powerful, and greatly interferes with the rapid reversal of currents in the surrounding coils, exercising, as it does, a powerful retarding influence.

CHAPTER V.

Electro-Magnets and Armatures.

WE have learned from the preceding chapters that an electro-magnet consists of a mass of iron in which magnetic polarity is raised by the circulation around it of an electric current, and that a large proportion of the energy of the current may be obtained mechanically in the form of a magnetic attraction and repulsion upon a second mass of iron or steel, called the armature.

In the most simple form of electro-magnetic motors the armature consists of a mass of soft iron attached to some kind of mechanical device, having for its object the conversion of a reciprocating into a rotatory motion. In other forms of the electro-motor the armature is also a magnet, being composed either of hard steel, in which case it is a common permanent magnet, or of soft iron, forming a second electro-magnet. It is common to call the fixed magnet the *field magnet*, and the moving one the armature.

The combinations which may thus be employed are very numerous. Electro-magnetic motive engines must, however, be either rotatory or reciprocatory in their action. The rotatory machines are of one kind, in which, however, numerous different combinations of simple armatures, permanent or electro-magnets may be utilised. The reciprocating motor must belong to one of two kinds. First, that in which the armature (of any type) is directly attracted to and alternately repelled from the poles of an electro-magnet. Secondly, that form in which the armature is represented by a plunger of iron, which is alternately attracted into and expelled from a hollow coil of wire or solenoid. In this case also both soft iron and magnetised steel armatures are used as plungers.

We have now to consider the influence or mechanical effect yielded by electro-magnetic devices of different types; and when these are variously arranged and excited by the currents. We are also concerned in the primary principles affecting their use as converters of electrical into mechanical energy.

Soft Iron, as used in the construction of electro-magnetic machines, signifies a kind of tough and carefully-annealed iron of good quality, *which allows of its being rapidly magnetised and demagnetised*. This kind of iron also receives stronger charges of magnetism than hard iron and steel—that is, it does not attain the condition known as saturation until its attractive or suspensive power is greater than the maximum force obtainable from equal bulks of hard iron or steel. This may readily be demonstrated by means of a magnetising coil and pieces of the kinds of iron and steel mentioned. What is generally known as Swedish iron is found to most readily answer the above conditions, but the best common iron of commerce is generally used. It is found that the oftener the cores of a magnet are magnetised the softer the iron becomes.

Annealing is generally effected by soaking the iron in a blood-red fire for some time and then cooling out *very slowly*—a condition most easily effected by allowing the fire to die out without removing the iron. Some authorities and iron-workers maintain that plunging the iron in water when it still retains considerable heat has a softening effect, but experience does not appear to verify this opinion, at least so far as magnetic ductility is concerned.

Steel, when employed in the construction of electro-magnetic machines as permanent magnets, should be of the best quality only. The metal should, while yet in its soft state, be forged, drilled, and otherwise finished in the form it is required to present. It is then to be hardened by bringing it to a blood-red heat in a slow fire and cooling out suddenly in water or oil. When thus hardened, and of good quality, a file should not cut the steel. Hardening is therefore the reverse of annealing. Hardened steel of good quality may be said to permanently retain magnetism, while softened iron most readily loses it. Care should be taken in hardening steel not to allow the temperature to rise above that degree indicated by a cherry-red colour, otherwise the quality may be seriously impaired by "burning."

Tempering is a modifying process, intended to make the steel rather less hard or brittle. It is frequently practised in the case of permanent magnets, in order to secure a greater suspensive power; but the tempering must not be carried too far, otherwise the steel will be restored to its soft condition, and the magnetism will not be retained.

When a piece of hardened and brightened steel is slowly heated, it begins to change colour by oxidation, and at 430° F. assumes a very pale straw tint. At 450° the tint is darker and more apparent. At 470° it is a dark straw yellow, and gradually deepens throughout all the grades of colour, down to a deep blue at 570°, which is known as " spring temper." It is found that good steel may be softened down to a straw tint (500°) and still permanently retain magnetism, while inferior qualities should not be raised in temperature beyond a pale straw tint.

Horseshoe magnets, or U's of steel, to be tempered must be brightened at several parts of the surface, to enable the operator to observe the degree of softness. The steel should be laid upon a thick plate of iron, previously made red hot, and narrowly watched until the required degree of softness has been attained, when it should be suddenly cooled out in water.

Magnetisation.—When ready for receiving magnetism, the steel may be subjected to the influence of a strong current of electricity passed through a coil of about two or three layers of insulated wire. Or it may be magnetised by being subjected to frictional contact, continuously in one direction, from the poles of a strong electro-magnet. In either case a soft iron armature should be kept across the poles of the new magnet. A permanent magnet of some strength may be used for the same purpose. The current process is decidedly the better. The current should be passed for a few seconds, and the circuit may be broken and completed several times during the process. This latter method aids the magnetisation.

Armature Iron should, as a rule, be as soft and ductile as possible. In some instances, however, when the armature is not to be subjected to reversals of polarity, or when it is required to act to a certain extent as a magnet, malleable cast and ordinary cast iron may be employed with advantage. The armature should, as a rule, be of a size and weight sufficient to fully engross the inductive effect in the magnetic field. Its sectional area should be as great as that of the magnet itself. Special instances arise, however, in which the armature should be light and of a smaller section than the core.

Electro-magnet Cores.—The core of a magnet may consist of a solid bar, a plate, a tube, or a bundle of wires. In ordinary cases it should be of the softest iron. Under ordinary conditions of magnetisation, when the cylindrical pole of a magnet is examined under the influence of the current, it is found that the magnetic attraction is exerted most strongly, not at the centre but towards the encircling edges. In fact, when the magnet is rapidly magnetised and demagnetised the central portion remains almost passive. This fact has led to the employment of cores of all conceivable shapes, among which tubes have found most favour, more especially where the polarity must be frequently reversed. When a magnet is subjected to the influence of a sufficiently powerful current, the whole of the core is magnetised to saturation, and under such conditions a solid-core magnet would be capable of exerting greater attractive power than a tubular-cored magnet.

In cases when the magnet must be rapidly alternately made and unmade, a solid core is a disadvantage, because it acts to a great extent as a magazine or condenser of the magnetic energy. Hence, the tubular and flat forms of core prove the most effective forms of core under these conditions. Further, for smaller apparatus than electro-motors, the tubular core may be split longitudinally, or a bundle of wires may be used as a core. Such magnets as these prove very effective under the influence of small currents.

In *length* the cores of electro-magnets of the common form should not be less than ten times the diameter, and in most cases should be much larger. The coils may be wound all over the bar, but it is usually found most convenient to arrange the wire upon two bobbins, rather shorter than the two limbs, and to join up the wire in the same manner as if it were wound all over the bar in a uniform direction or unbroken helix.

When the core is composed of one bar of iron the most effective form it can assume is that known as U shape. The distance between the limbs is to a certain extent regulated by the amount of wire to be coiled upon each limb. It should seldom be greater than four times the diameter.

It is found by experiment that enclosing the limbs of a magnet of this type by a pair of brass or copper tubes exercises a considerable effect upon the attractive force. Some authorities speak of a gain of 20 per cent.

Another construction of electro-magnets offers some advantages, because the attractive force for a given current is increased. The core is a tube, around which coils of wire are wound as usual. Over this another iron tube and coils are placed, thus concentrating a great portion of the inductive effect within the magnet itself. The construction is however expensive and troublesome.

There is obviously no limit to the number of different devices which may be resorted to in connection with cores for electro-magnets, but the question finally turns upon the consideration whether the gain in force equals the extra first expense, and the disadvantages of complicated methods of construction.

The core of a U magnet need not necessarily be in one unbroken length. On the contrary, almost every magnet core employed in telegraphic and other apparatus is made in three sections—two limbs and a "yoke" or junction piece—to which the limbs are screwed. In extension of the sectional method of construction the field magnets employed in all the more notable dynamo-electric machines and electro-motors are constructed of several pieces connected together with bolts and nuts. The construction of the whole machine is thus much simplified, and the loss of force at junctions inappreciable. All contact faces should, however, be both clean and flat, because the magnetic circuit continuity is very easily severed. The effect of securing the continuity is to make each limb into a separate magnet with two poles, or a total of four poles, reducing the efficiency of the magnet to one half

Extension of the Poles.—By attaching extension pieces of soft or cast iron to the extremities of a magnet, its poles may thereby be extended outwards from the core itself. The electro-magnets used in the Gramme and other dynamo-electric machines are extended in this manner to embrace the revolving armature.

Electro-magnetic Solenoids.—When a current circulates in a coil of wire which forms a hollow cylinder, a plunger of iron or steel will be attracted or sucked into the chamber. When the plunger is a permanent magnet it is expelled again upon reversing the direction of the circuit in the solenoid. Another method is to employ a helix, the lower half of which contains iron, while into the upper half a plunger of iron is attracted. A very considerable yield of mechanical energy from a given current may be obtained from an electro-motor arranged according to these principles.

Conducting Wires for Electro-magnetic Apparatus. Copper wire is almost invariably employed in the construction of electro-magnetic machines. It is usually insulated by a covering of silk, cotton, hemp, or guttapercha. The insulating covering is applied to the wire by a special covering machine. Wires insulated with silk occupy less space than those covered with cotton, and the insulation is more effective. It is of much importance to use only wire of good conductivity, and carefully protected, so as to reduce the resistance of the apparatus as far as possible and insure good insulation. For currents of low electro-motive force the covering may be thinner than for currents of great force.

The following list contains all the sizes of insulated copper wire in common use. The finer sizes from 20 upwards are included, although only employed for the lighter kinds of electrical apparatus.

LIST OF INSULATED WIRES.

Birmingham Wire Gauge.	Size in Millimetres.	Length. Feet per lb.	Resistance. Pure Copper Ft. per Ohm.
No. —¼in. diam.	6·37 diam.	5·29	6046·5
6	5·08	8·26	3869·8
8	4·31	11·18	2861·0
10	3·55	16·87	1896·2
11 ⅛ in.	3·17	21·16	1511·6
12	2·79	27·32	1170·6
13	2·41	36·63	873·1
14	2·15	46·28	691·0
15	1·92	58·77	544·2
16	1·651	78·24	408·8
17 1/16 in. (nearly)	1·440	101·75	314·3
18	1·274	132·22	241·9
19	1·143	163·25	196·0
20	1·016	206·60	154·8
22	·813	322·81	99·1
24	·635	528·90	60·5
26	·483	915·78	34·9
28	·406	1291·3	24·8
30	·355	1690·4	19·0
31	·305	2295·6	13·9
32	·254	3305·6	9·7
33	·250	3586·8	...
34	·244	...	8·9
35	·221	4367·3	7·3
36	·200	5296·6	6·0
37	·170	7363·7	4·3
38	·147	9826·4	3·3
39	·106	18739·0	1·71
40	·099	21732·0	1·47

Of the sizes given in this table, No. 6 may be used for the strongest currents. The sizes from No. 6 to 14 are commonly employed in dynamo-electric and electro-dynamic engines. The column giving feet per ohm is only correct for pure copper, at 60° F. From 5 to 15 per cent. should be added for ordinary copper. It is important not to employ wires too large in the construction of electro-magnets. The average effective number of *layers* of wire upon ordinary electro-magnetic machines is 4. In any case this number, or any greater number, should not present a total thickness of more than half the diameter of the core, when the core is cylindrical. When the core is flat, as in the Siemens electro-dynamic machines, the exciting layers of wire frequently exceed it in thickness. Consistent with efficiency, the resistance of the circuit should be kept as low as possible; but this must be intelligently understood. The leading wires, or those having no inductive effect upon any portion of the apparatus, should be as large as convenient. It is usual to employ flexible *cables*, composed of several small wires twisted into a rope, for this purpose. These are usually so large that their resistance, for short lengths, may be left out of account. But in the coils, the influence of the absorption of energy by numerous turns of wire must not be overlooked. The efficiency of a given electro-motor may, under certain circumstances, be immensely increased by doubling the number of turns on the magnets; or it may be immensely diminished, according to the internal resistance of the electric source. As a general rule, the motor coils should present resistance at least equal to that of the electric source.

From these considerations the following general conclusioms have been deduced. For electro-magnetic portions of machines and motors weighing from 20lb. to 50lb., as much of the sizes of wire from No. 8 to No. 14 may be wound on until the resistance equals that of the electric source to be employed. As a rule, the largest gauge of wire is used upon the largest machines. This implies, generally, that the largest wire should be used for a very low interior resistance in the electric source. Also, that not *less* than three layers should be employed, and this in the most advantageous position. Also, that the sum of all the resistances shall be greater, rather than smaller, than that of the electric source.

When the resistance of the electric source is considerable, say over 4 or 5 ohms, a finer size of wire than usual should be used, since in such cases four layers of fine wire will prove more effective in developing magnetism than six layers of a larger size of wire. When it is calculated that a given size, say No. 12, will, when wound in the necessary number of layers, yield too high a resistance, a larger size should be employed. Hence, the size of wire is controlled both by the number of layers necessary to obtain maximum magnetisation and by the resistance at the electric source.

The rules already given governing the construction of electro-magnetic arrangements, indicate that the magnetic attraction depends upon the number of convolutions; but this must be considered in relation to the fact that, beyond a certain distance from the core, the influence of each layer of convolutions is increasingly less. It is therefore possible to overdo the number of convolutions under the impression that a gain will thereby be secured. The common expression that the strength of a magnet increases as the number of turns can be true only within certain limits, which must be determined for each specific case.

In cases where the armature of the magnet is intended to be magnetised and demagnetised alternately and rapidly, the number of layers in the exciting coil should seldom be more than four. This usually necessitates the use of comparatively fine wire, which may be made to describe a considerable number of turns in a small space. Fine wire will in this case prove more effective than thick wire, the bulk of which might necessitate the winding of five or six layers.

It is extremely difficult, however, to lay down, even approximately, correct rules for the coiling of electro-magnetic arrangements. The electromotive force developed at the electric source, the total resistance, the nature of the generator, the velocity of the motor, and other considerations, must be taken into account. When the leading wires are long or thin they should be counted into the resistance of the motor.

When a given magnetic force is required to be excited by an electro-magnet, different electromotive forces will be required to produce it, according to the conditions of the circuit. Thus, a magnet that acquires a certain attractive force with a constant current in one direction, will not yield the same effect with an interrupted current of equal strength. It is found that to magnetise and demagnetise rapidly an electro-magnetic bar involves a large expenditure of electrical energy, probably three times that necessary to maintain the same magnetic effect from a current constantly flowing. A back electro-motive force is produced at each change, and the waste of energy is manifested in the coils and cores as heat.

From these considerations it is evident that any of the regular statements assumed to express the laws governing electro-magnets are frequently delusive or misleading, since the conditions are continually changing, each change recording its effect upon the resistance of the circuit. But the whole question should be viewed broadly, with its various correlations, for specific cases. A change, whatever its nature may be, implies either an expenditure of additional energy, or a diminution in the resistance. When the current is uninterrupted the expenditure of energy, when the magnet does no work, is clearly, after the first instant, that due to the wire resistance alone. But if the magnet does work, or acts in the maintenance of a magnetic field, which is utilised for a useful purpose, there is an additional expenditure due to " work " pure and simple, or to counter electro-motive force. This counter electro-motive force, due to the reaction of the magnetic force in the field when doing work, is often erroneously regarded as increased resistance, analogous to simple conductor resistance.

Again, when the current is frequently interrupted, the magnet will do less work, and yet the expenditure of energy will probably be as great as before, a source of loss which may clearly be traced to absorption or conversion of energy in impulsions throughout the circuit—impulsions which are only to be recoverable as *heat*, which in this case is useless. In short, any addition to the expenditure of energy upon a magnet which cannot be recovered as work may be assumed to be converted into heat in the wires and cores.

Efficiency of Electro-magnetic Apparatus.—An

electro-magnet is known to be an effective convertor of electric energy into work by the following signs: When the magnet is inactive—that is, removed from armatures or other pieces of iron, its resistance should be only a little greater than that of the length of wire in the coils. When the magnet sets an armature in motion, so as to do useful work to its maximum power, its resistance should appear to increase very considerably—due to back electro-motive force—that is, the current flowing will diminish. Hence, it is correct to assume that *the proportion the extra resistance* (diminution of current) *bears to the total is the measure of that portion of the expended force converted into work.* The greater this proportion the more effective the motor, and the less costly its working.

CHAPTER VI.

Electric Accumulators or Magazines.

When an electric current is passed through a cell, composed of two plates of the same metal plunged in water, a result ensues which may be assumed to be the storing up of a certain quantity of the electric energy. That an accumulation of force has actually taken place may be demonstrated by disconnecting the cell and closing its circuit through a galvanometer. A powerful rush of current is at once indicated, the direction of which is opposite to that of the primary current. When the plates are examined, while still in a condition of electrical accumulation, it is found that a portion of the water has been decomposed, the constituent gases being deposited in a layer or cushion upon the two plates. The oxygen appears upon the positive plate, or that through which the current enters the cell, and the hydrogen appears upon the negative plate, or that by which the current leaves the cell. The hydrogen film is electro-positive to the oxygen film. A certain amount of energy is charged or conferred upon the molecules, so that, when the plates are connected by a wire, a current passes from the hydrogen to the oxygen. The result is a recomposition of the water; the gases reunite, and the energy conferred upon the molecules is again set free in a current reversed to the charging current.

A German physicist named Ritter was the first to actually construct a battery from which these secondary currents could be obtained. Volta and Bicquerel showed that the secondary action was in no way dependent upon the plates, but that it arose simply from the layers of electrically-charged gases upon the plates, and that the plates simply acted as conductors or surfaces upon which the gases might be deposited. Grove constructed a "gas battery," in which oxygen and hydrogen were employed in this way to furnish a current, both the elements being of platinum.

When the water is acidified its conductivity is increased, and the accumulating tendency is stronger. Since the effects obtainable in this way from platinum, silver, or gold plates were of a feeble and limited nature, efforts were made by several physicists to produce a combination capable of storing a larger amount of energy. It was discovered that not merely gases but also salts and other substances yielded the reactionary currents. M. Planté, of Paris, discovered in 1859 that when two sheets of lead were plunged in acidified water they yielded effects more powerful and lasting than those from platinum. When the charging current was continued for a sufficient time a peroxide of lead was formed upon the oxygenised plate, and hydrogen was yielded or set free at the negative plate. It was found that when this stage was attained a very great quantity of electricity was stored by the cell. When the plates were connected by a wire a continuous current flowed. M. Planté found also that, in the course of the yielding up of the energy, the oxidised plate became deoxidised, and that its oxygen passed through the liquid and attacked the opposite plate, oxidising it. The plate which had thus been oxidised and deoxidised presented the appearance of spongy lead, presenting an enormous extent of surface to the action of the current. Each time the cell was used, its capacity for storing electricity was augmented, since the surface available was increased. Thus, one of the plates was always in a condition of partial oxidation.

These observations led M. Planté to greatly extend the surfaces of the lead plates. He placed

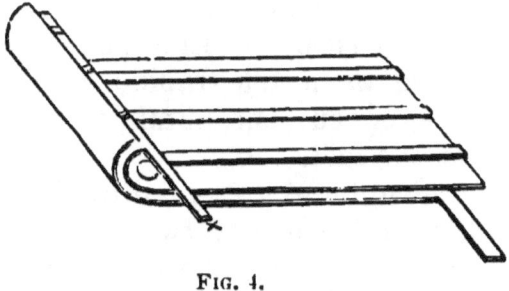

Fig. 4.

canvas between two large thin plates, and rolled them up in a close spiral. The whole was then plunged in a vessel containing acidulated water. This form of the accumulator cell was found to possess enormous storing capacity, and it improved after each period of use.

Subsequently, it was found that, in order to obtain the full effect, a free small space should be provided between the plates for the escape of gases. This led to the construction of the cell finally adopted by M. Planté. He laid upon a large sheet of lead a series of three india-rubber bands, placed at equal distances apart. Upon this was laid the second lead plate, and the two were rolled up in a close spiral, placed in a deep vessel, and moistened with acidulated water. By means of a large cell of this description, charged by one or two small voltaic cells, an accumulation of electricity capable of rendering incandescent several inches of platinum wire was produced. The current lasted for a period depending upon the condition of the cell, or from ten minutes to one or two hours. Thus, one great advantage of the Planté cell consisted in the facility with which the accumulated energy could be set free for practical purposes. It did not come off with a rush, or in a momentary spark, as in the case of a telegraphic condenser. It flowed in an even current, diminishing, however, somewhat at the last.

Another advantage presented by the secondary cell was that a battery of twenty cells might be charged by two small voltaic generators, at a tension of two volts, the whole battery being arranged as one large cell. When charged, the connections could be changed by a commutator, so as to make the secondary cells into a battery of twenty cells in series. By these means a current having a force of thirty or forty volts might be obtained. It is true that, as thus arranged, the secondary cells became exhausted more rapidly than a single cell of large surface, but the advantage of being able to obtain the force of twenty cells by means of a current from two only was very great.

The effects obtainable from Planté's battery or accumulator were not sufficiently lasting to serve many practical purposes, although many experiments, involving the production of powerful currents, were performed by means of the type of cell just described.

M. Faure, in the spring of 1881, effected a very great practical advance in the means employed for accumulating electrical energy. Acting upon the hint given by the fact that peroxide of lead was formed upen the positive plate of Planté's cell, he tried the effect of coating the plates direct with red lead. The peroxide was mixed into a thick batter with water, and distributed in a thick cushion on the lead plates, which were then separated by cloth or felt, and rolled up in a close spiral, as in Planté's cell. The result was not only a combination which was at once ready to receive a charge, but an arrangement which possessed about thirty times the accumulative power of a Planté cell.

The method of construction adopted by M. Faure, or by the company who possess the rights of trading in the invention, is as follows: Two sheets of lead are taken, 6in. wide, one being 20in. long, the other 16in. long. Thick leadfoil is used for this purpose, the larger sheet being the thicker. Each sheet is furnished with a long strip of lead, projecting from its side, to form the electrical connection. Each sheet is then coated with a layer of red lead, made into a thick paste with acidulated water. A piece of parchment, or parchment paper, is laid upon each of the treated surfaces, and the other sides are similarly coated and covered with parchment. Each sheet is then sheathed in thick felt. The longer sheet is now taken, and the smaller placed above it, and the whole is rolled up together in a tight spiral. It is thought advisable to place several strips of indiarubber obliquely between the sheets, and interiorly of the inner sheet, in the process of rolling up, so as to afford a means of escape for the gas, if necessary. This spiral is now introduced into a containing vessel or jar of lead, strengthened by bands of copper. The interior of this jar is also coated with red lead and felt, to increase the active surface. The conductor strip belonging to the outer sheet of lead is now soldered to the containing vessel, which is filled up with water, containing about one-tenth of sulphuric acid. The electrical connections are made by attaching terminals to the remaining lead strip and to the containing vessel.*

A cell of this description, when charged, by being connected with two or three weak telegraph cells (Daniell type) for one hour, will accumulate sufficient electric energy to heat an inch or two of fine platinum wire to redness, and to furnish a current equal to that from ten or twelve telegraph cells for the space of fifteen minutes. When fully charged, by being connected with a more powerful battery, or a dynamo-electric machine for a few minutes, it will yield a powerful current for several hours afterwards. The time during

---

* In the more recent construction of this accumulator cell the shape is oblong.

which the cell may be allowed to remain in a condition of accumulation appears to vary, and does not appear to have been ascertained. Its force falls off, probably by leakage between the plates. By means of better insulation, it is probable that the energy stored might be kept unaltered in quantity for months. A cell made in the common manner will, however, retain its charge almost unimpaired for many hours, and in some cases days have elapsed before the stored electric energy was used.

It is therefore evident that by these means it is practicable to store up a considerable amount of electric energy in a small compass, or, indeed, any required amount, according to the surface of the lead plates, and to convey it from place to place by the ordinary means. There is no possible danger in carrying about energy in this form. It is available for use at any moment by simply providing conducting wires, connected to the terminals of the electro-motor or lamp.

By a calculation and experiments made by M. Reynier, afterwards verified by Sir William Thomson, it was shown that a Faure accumulator weighing 165lb. can store, and afterwards yield up, energy to the extent of 2,000,000 foot-pounds, or 1hp. maintained for the period of one hour. According to the observations of Sir William Thomson, the loss of energy involved in the charging and discharging of the elements would not amount to more than 10 per cent. of the total energy. If the process of charging is pushed on too rapidly, there is a loss, owing to actual conduction of energy through the cell of a certain portion of the energy not effective in changing the condition of the plates. This loss will in most instances appear as heat. In most cases it is advisable to prolong the charging process, but there is a limit to this. The work may be done too slowly, and partial local action may occasion a slight loss, but in any process not uselessly slow there will be very little loss from this cause.

According to the experiments of Sir William Thomson, with a view to determining the efficiency of these accumulators, and their applicability to the purpose of driving tramway cars through the medium of electric engines, it would appear that an eighth of a ton of accumulators would work very economically for six hours at one-sixth of a horse power. It would work much less economically for one hour at 1hp., but not so uneconomically as to be practically fatal to the proposed use. He considered that it is very probable that a tramcar arranged to take in, say, $7\frac{1}{2}$cwt. of freshly-charged accumulators, on leaving headquarters for an hour's run, may be driven more economically by the electric energy operating through an electro-dynamic machine than by horses.

There can be no doubt that, under ordinary conditions, the method adopted in the electric railways of transmitting the current along the whole course of the line, through the rails, or through a special rail, will offer greater advantages than that of carrying accumulators of this type in the car or train itself. But in many instances it is not practicable to lay down or use special or other conductors, and in such cases the employment of the accumulators would prove best suited to the purpose of electrically propelling tramcars in the highways of towns.

In the process of charging those cells, it will usually prove most economical to arrange all the plates in the manner of one large cell. Thus, if there be three Faure cells, let all the outside leads be connected together, and all the inside leads together also. From these two sets two wires mnst be led to the electric source, and the primary current passed.

When the energy in the accumulator is required for use, it may either be drawn under a potential of about 1·9 volt, as a current of large quantity but low electro-motive force, or it can be drawn under a potential of about 5 volts by connecting the three cells in series, as follows: The inner lead of No. 1 cell to the outer lead of No. 2, and the inner lead of No. 2 to the outer lead of No. 3, carrying wires from the extreme leads at either end to the work. In the first case, the current will prove effective against a small resistance only. It will not prove effective in a circuit composed of thin wire. In the second case, the interpolar resistance should be considerably greater. The internal resistance of the cell or battery must, in most cases, be used as a guide with regard to the extreme resistance to oppose to it. A Faure cell, exposing a total surface of lead of 10 square feet, will frequently show an internal resistance of nearly 0·5 ohm. Thus, in the case of three cells, as in the above instance, the total resistance, when arranged in series, would be about 1·5 ohm, to which might be opposed, with advantage, an interpolar resistance of not less than 2 ohms.

The type of secondary cell exhibited in Fig. 5 may be employed with advantage for experimental

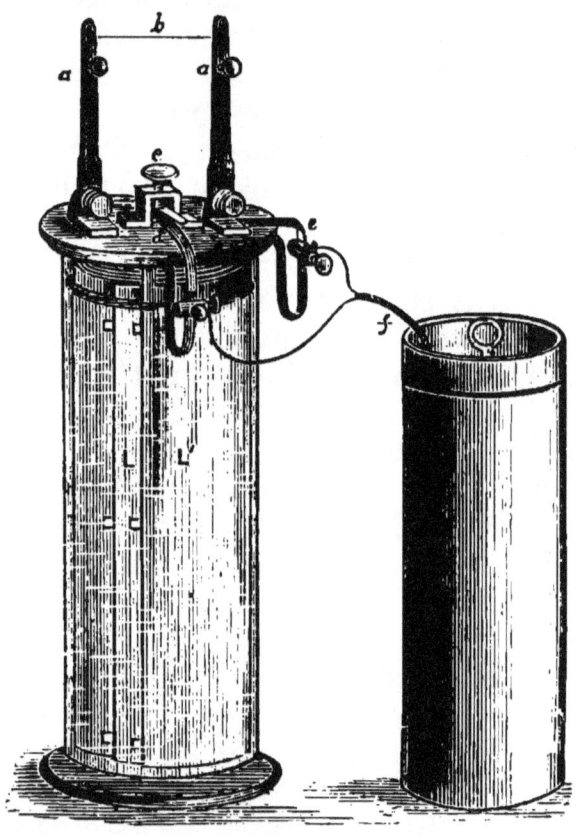

Fig. 5.

purposes. It is arranged to present a total surface of lead of 10 square feet. The plates and lead oxide are arranged in the manner already described, and two strips of copper, firmly riveted to the plates, lead out of the cell to form the connections. The upright pillars *a a* represent a pair of brass clips, connected to the two poles, by means of which, when required, a fine platinum wire may be stretched across, by the incandescence of which the progress of the charging may be observed. The central depressing screw closes the circuit of the cell through the platinum test wire, when required.

This cell, for experimental purposes, may be charged most easily by a voltaic cell consuming zinc. The author has employed for this purpose a type of cell which he has devised and calls the sustaining battery, because it is arranged to sustain its vigour for any required length of time. A description of one element of the battery will be found in Chapter IX. It is connected to the secondary or accumulator cell in the ordinary way. One cell of the sustaining battery, or a Bunsen cell, or even a few Daniells may be used to charge the accumulators, even if they present a total surface of 100 square feet. Two cells are more effective.

The potential of the charging battery or dynamo-electric machine has but little influence on the electro-motive force of the accumulator. It will remain rather under 2 volts per cell, whether the charging source be of 2 or 50 volts potential.

CHAPTER VII.

Construction and Efficiency of Electro-Motive Machines.

The fundamental laws governing the production of motion by electrical agency, regarded from a practical point of view, are severally treated in the second chapter. We now enter upon a consideration of the most suitable means, within present knowledge, of turning the dynamical properties of electricity to useful account.

A vast amount of skill, ingenuity, and labour has been expended in almost fruitless endeavours to construct efficient electro-motors. This is true of the period succeeding the discovery of a means of furnishing a practically constant current of electricity, but preceding the discoveries of the past fourteen years, upon which the great modern developments are based.

A large proportion of former abortive attempts to produce useful electro-motors was conducted upon erroneous notions regarding electricity, and by inventors possessing all the requisite knowledge save that most essential to the success of their experiments. The patent records of many countries testify to the numerous failures incurred by persons clinging tenaciously to the hopeless idea that a zinc motor could compete with even a very indifferent coal motor. Further, it is now evident that even the electro-motors proposed, although exhibiting every mark of the highest ingenuity, were generally constructed upon wrong principles, frequently indeed in utter ignorance of the elements of electro-magnetic science. At the same time it may be considered that no class of natural phenomena is so apt to lead the inventor astray as those involved in the production of mechanical movements from electro-magnetism.

It is only a few years since the grandest and most fruitful doctrine of modern science, the conservation of energy, revealed the fact that zinc never can compete with coal in the production of large moving powers. With the publication of Dr. Joule's deductions, in which zinc was shown to possess a definite potential energy, considerably less than the potential energy of coal, fell the hopes of all those who had borne up against repeated failures. This disposed of the idea that a zinc motor ever could compete with steam, on any considerable scale.

But the gradual growth of an exactly opposite branch of electrical science brought new facts to light. The magneto-electric machines of Siemens and others led the way to the production of the most powerful currents, from developments of the same principles, by Holmes, Nollet, Wylde, and others. Profound scientific knowledge, and more especially an extended familiarity with electricity and its correlatives, were brought to bear upon the question of the production of *electricity from mechanical motion*. The question was at the time of more interest and importance than that of producing motion by electrical agency, and it promised to open up a wider and more fruitful field than that in which so many inventors had already failed.

It was not, however, until about the year 1872 that the striking result of passing a current *into* a dynamo-electric machine was observed. M. Gramme was one of the first to discover that his well-known dynamo-electric *generator* formed an equally effective *electro-motor*.

The late eminent Professor Clerk Maxwell considered this perhaps one of the most important and interesting discoveries made within the preceding ten years.

It was speedily discovered, upon reasoning the whole question of generating magneto electricity as it were backwards, or upon a converse basis, that an efficient dynamo-electric generator must in most instances form also an effective electro-motor. The electric and magnetic conditions are identical. This singular coincidence attracted much attention, but the development of the principle, or its application to useful purposes, has been considerably delayed by the counter attractions of electric lighting.

Most of the dynamo-electric machines now used may also be employed as motors. Siemens' machine has been more extensively used as a motive engine than any other. Examples of its application to various purposes are offered in another section of the work. Gramme's, Brush's Maxim's, and many other types of dynamo-electric machines have likewise been applied in the same way, and will probably shortly occupy an important position as motive engines. The construction of these machines is treated upon at considerable length in works devoted to the subject of electric lighting, and they have been described very ably in various papers read before the scientific and engineering societies of Great Britain. The subject does not on this account enter into the scope of the present treatise. But examples of electro-motors, specially designed for the conversion of electric into mechanical energy, are offered, in order to render the section devoted to construction as complete as possible in itself. One or two of these examples are but models on a small scale, but they serve to demonstrate the various methods of construction adopted, or which may be adopted, in the production of machines of any required dimensions. Details of the action of a model, as employed by scientific lecturers for the elucidation of the principles of those machines are also offered, in order to carry out the various electrical and magnetic principles already expressed. Meantime it may be mentioned that the types of construction resorted to prior to Gramme's discovery have been entirely abandoned, and the new motors are designed closely upon the lines of successful dynamo-electric machines.

With reference to the efficiency of dynamo-electric machines when employed as motors, the author is enabled, through the kindness and liberality of Messrs. Siemens Bros. and Co., to place the following very instructive and valuable particulars before the reader. They relate to the Siemens machines.

1. In effecting the transmission of power by

electricity, one dynamo-electric machine is driven by a strap or gearing in the ordinary manner from a steam engine or other motor. This machine generates a current of electricity, which is caused to pass through leading wires, of the required length, to a second and precisely similar machine, placed some distance off. Thus the first machine generates the current, which, passing through the second machine, causes it to revolve, and renders it capable of exerting mechanical energy.

2. The amount of work done by the second or motor machine is a certain percentage of energy employed in driving the first. If the speed of the first machine is kept constant, this percentage varies with the speed of the second, and is at a maximum when the speed of the second machine is about $\frac{1}{2}$ that of the first. As an illustration of this, the following are the results of an experiment with two Siemens' small machines, one as first and one as second machine. The first machine ran with a constant speed of 1,100 revolutions per minute.

Number of Revolutions per Minute of Second or Motor Machine.	Percentage Power Reclaimed by Second Machine.
884	34
808	43
767	44
625	45
181	39
385	32

3. The amount of energy actually reclaimed averages, under favourable circumstances, between 40 and 60 per cent. of that exerted by the first machine. Keeping the relative speeds of the two machines the same, this percentage rises with the increase of speed of the first machine, its most

favourable velocity being the greatest the machine will bear without becoming excessively hot. The following is the result of an experiment with two of Siemens' medium sized dynamo-electric machines, as an example:—

Number of Revs. per Min. of 1st Machine.	Ditto of 2nd Machine.	Current in Webers.	Percentage of H.P. Reclaimed by 2nd Machine.
693	395	21·0	40·7
960	520	28·2	46·1
1155	620	31·9	47·7

The electrical resistance of the two conductors joining the terminals of the two machines was equal to 0·325 Siemens's unit. (Siemens's unit = 0·9563 ohm.).

A Siemens medium machine is capable of working up to about 2½ actual horse power (as an electro-motor), the energy expended upon the machine furnishing the current being from 4 to 5 horse power.

The introduction of conductors of greater length and resistance to connect the machines together reduces the useful effect in the ratio of the amount of resistance introduced. Two of Siemens' small machines gave the fellowing result, the velocity of the first machine being kept constant at 1,100 revolutions per minute, the resistance of the dynamometer brake being also maintained the same during the four trials:—

Resistance of Conductors in Siemens's Units.	Speed of 2nd Machine in Revolutions per Minute.	Percentage Power Reclaimed by Second Machine.
0	685	44
0·5	589	38
1·0	502	32
1·5	433	26

Distinction between Magneto and Dynamo Electric Machines.—It will be well, before proceeding further, to define the difference between the older magneto-electric machines and the more modern dynamo-electric apparatus.

The terms magneto-electric generator is generally employed to signify a machine in which the armature is caused to rotate in a field created either directly by a battery of permanent magnets or by an electro-magnet excited by the current from a smaller magneto-electric machine, specially reserved for that purpose, and forming part of the machine itself. This latter device is adapted in one form of Wylde's machines.

The term dynamo-electric generator is generally reserved for that form of machine in which an electro-magnet, *excited directly by the current of the machine itself*, is employed to set up the magnetic field. The current created by the first few turns of the armature in the weak residual magnetic field passes through the field magnet and strengthens the field, which in turn reacts upon the armature, which yields still stronger currents. Thus the action goes on, accumulating force at the expense of the driving power, until the field magnet is saturated or the current fails to increase by augmenting the velocity. A current yielding energy equal to about 90 per cent. of the driving power is thus available for external work, the armature, field magnet, and work being all in one circuit. Sometimes the field magnet is excited by an independent machine of small size. This plan is adopted in cases where great steadiness in the currents is required, as in some instances of the production of electric light.

From these considerations it is evident that, as soft iron can manifest magnetic power enormously greater than that of steel, the electro-magnetic field machines must necessarily be vastly more powerful, size for size, than any possible magneto-electric machine.

And conversely, when regarded as electro-motors, it is clear that the dynamo-electric apparatus must be capable of converting much larger currents than magneto-electric machines.

Dynamo-Electric (Primary) Machines Connected in Parallel Circuits.—When two dynamo-electric machines are connected together in parallel circuit, *i.e.*, both positive poles to one leading wire, and both negative poles to the other wire, the current obtained is commonly about 20 per cent. stronger than the current produced by the machines working separately, or in separate circuits. This property of the current-generating machine is taken advantage of in the primary arrangements used in the working of electric railways.

The necessary connections are usually arranged between the two machines direct, and the leading wires are in this case connected to the c and z poles of one of the machines. The connections between the two machines are as follow: C of first machine is connected with c of the second machine; z of the first to z of the second; m (the electro-magnet terminal on the z side of the first machine) to b (terminal on brush, on z side of second machine). B of the first machine (brush terminal on z side) to m (magnet terminal on z side of second machine). M^1 (magnet terminal on c side of first machine) to b^1 (brush terminal on c side of second machine).

B^1 (brush terminal on c side of first machine) to m^1 (magnet terminal on c side of second machine). The leading wires are connected to c and z of the second machine. These connections will be rendered clear by reference to the engraving of the Siemens machine, Fig. 6.

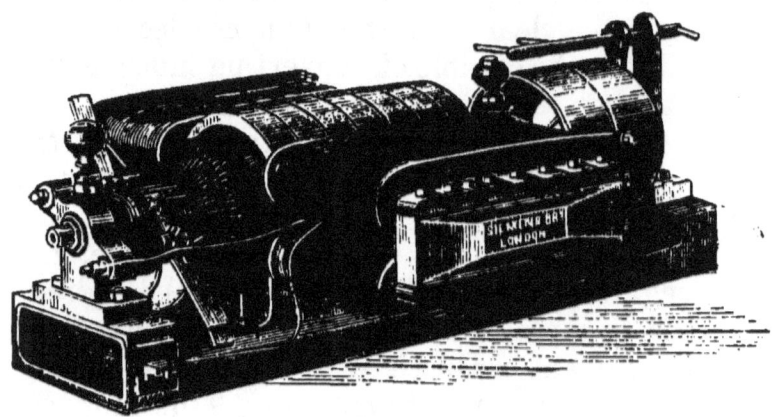

Fig. 6.

Use of Dynamo-Electric Apparatus as Current Generators.—Like voltaic generators, dynamo-electric machines must, to produce the maximum of effect, be arranged according to certain conditions. The resistance should be the first consideration. The internal resistance of the machine should in all cases be known. When the leading wires are short, their resistance may be left out of account, and that of the electro-motor only calculated. In cases where the current-generating machine is so arranged that its field magnet is in the working circuit, it must be observed that any increase in the external resistance will considerably affect the influence exerted by the field magnet, and will thus rapidly reduce the current produced.

But in cases where the field magnet is excited by a portion of the current through a parallel or *shunt* circuit, so that the greater amount of the current may choose the lesser resisting path, an increase in the external resistance, or working circuit, will really augment the electro-motive force of the machine, because a large proportion of the current will pass around the field magnet, increasing the intensity of the field. If the shunt and the external circuit were made exactly equal, they would resist equally, and the current flowing from the machine would be exactly divided between them. If the shunt (electro-magnet) circuit were made longer, or of greater resistance, a larger proportion of the current evolved would be passed through the working circuit. If the external resistance only be made to vary (as is generally the case, the field magnet circuit being always the same) the current in the field magnet will vary to a like extent. If the external resistance be increased, the field magnet would receive a larger share. If the external resistance be diminished, the field magnet would receive a smaller share. Therefore, by this method of shunting a portion of the current around the field magnet, the machine will produce strong currents when the external resistance is great and *vice versa*. So that changes in the external resistance carry with them, to a certain extent, their own remedy, and everything tends to produce uniformity of current. This method is more generally applicable in the case of electric lighting circuits, where a steady current is required, than to the case of electro-motors. But it is probable that it might be used with advantage in cases of varying external resistance, even with motors, as in the case of electric railways.

Precautions against "Short - Circuiting."—It is important to observe that the dynamo-electric machine shall not, by accident or design, be allowed to work through *too small* a resistance. When a highly-efficient machine is put upon short circuit, the current produced will be very powerful, and its quantity will rapidly increase, until the circuit wires become red hot, by reason of the energy being expended as heat in the machine. In such cases the insulating covering of the wires may be burned off and the machine rendered useless, until it is rewound with fresh wire. Before the wires begin to develope heat enough to burn the insulating covering, long and bright sparks will be observed to leap from segment to segment of the commutator cylinder.

Commutator Brushes or Collectors.—These are brushes of hard copper wire or thin sheet copper, fastened in brackets and made to press upon the surface of the commutator or collecting cylinder revolving with the axis of dynamo-electric machines. In most cases they are attached to holders connected together at diametrically opposite points of the commutator, so that any movement of one brush may be communicated to the other. By these means the relative positions of the collectors (at diametrically opposite points) are maintained. In the course of time the collectors, through the combined influence of friction and sparking, become worn, and they then call for readjustment. In such cases the ends of the brushes should be cut off square. The pressure applied at the point of contact must be sufficient to cause perfect continuity, but it should not be so great as to cause excessive friction. When the machine is set in motion, it should be observed whether force is lost by sparking at the brushes. When the sparks are large, they indicate that the point of contact is not correct in relation to the speed of the machine, or that the friction is too slight. The holder containing the brushes may usually be caused to rotate upon the axis, by which means it is easy to determine the best position for the brushes. The most advantageous point will be indicated by the absence of sparks, it is commonly a little in advance of the theoretical line of contact. The commutator drum should be kept freely oiled, to reduce friction; grit and copper dust should not be allowed to accumulate upon it. When worn, the copper contact segments on the drum may be replaced by unscrewing the armature wires from their inner ends. A spare set of segments usually accompanies each machine. In again fastening up the extremities of the armature wires, care should be taken to connect each to its proper segment, and to insure that the metallic contact is perfect. There must be no metallic connection between the segments.

The Principles of Motion in Electro-motive Engines.—Fig. 7 represents a model of an electro-magnetic motor intended to illustrate the principle of action in these machines on a practical scale.

The model illustrates the arrangement of parts intended to utilise both the attractive and repulsive effects of magnetism. In this instance the field magnet is represented as of the permanent type, and the armature as an electro-magnet, both of the U form, which, however, is not used in machines on a useful scale.

N and S are the extremities of a permanent steel magnet, *s* and *n* is a bar of soft iron, curved to the U form to bring its poles within the influence of the permanent or field magnet. Upon each

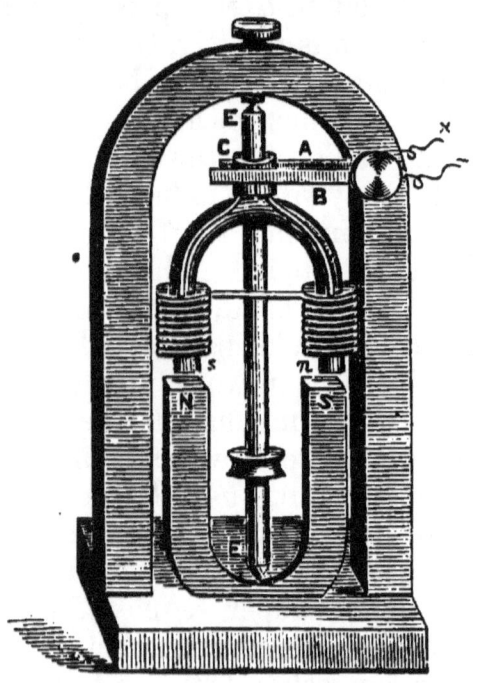

FIG. 7.

limb of this armature, as it will now be called, are wound two or three layers of insulated wire, the extremities of which lead to a *commutator* or current reverser. The armature and its commutator are rigidly mounted upon a vertical axis, capable of free rotation. The commutator is composed, first, of a cylinder of wood fitting tightly upon the axis. Two half rings of copper embrace this cylinder at opposite diameters, and to these half rings the extremities of the exciting coil are connected. As the commutator rotates, two springs A and B press upon it at diametrically opposite points. These contact springs convey to the commutator the driving current, which enters at the terminals + and —

Fig. 8 represents this current-reversing device more plainly. A and B are the springs conveying the actuating current; C the wooden or ebonite cylinder, with its two half rings of copper, the ends of which do not come into contact; 1 and 2 are

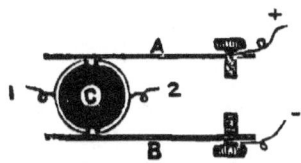

FIG. 8.

the extremities of the armature coil connected to the half rings.

The action of the machine is as follows: N and S being fixed, and s and n free to rotate, the current passes from the spring marked + around the armature, magnetise the armature, and returns to the electric source by the spring marked -

Assuming the armature poles to be near to the field-magnet pole, the current produces s magnetism in the pole opposite to N, and n magnetism in that opposite to S. The result is that the fixed magnet attracts both poles of the armature, which begins to rotate until the four poles coincide. All action would cease here were the current allowed to flow in the same direction. But just as the armature passes the central points of the fixed poles, *the half rings of the commutator exchange springs, and the current is reversed.* Hence the polarity of the armature is reversed, its n and s poles being now opposite to N and S. Since like polarities repel each other, it is clear that repulsion ensues at both poles, and the armature is driven forward through one quarter of a revolution. The current still continues to flow in the same direction, and the rotation brings the armature poles again within the influence of the fixed magnet, but in this case n approaches S, and s approaches N. Thus attraction ensues until the poles again coincide at the instant when the current is again reversed by the commutator, and repulsion takes place once more. Thus at each half revolution the current is reversed by the commutator, and attraction and repulsion take place in consequence. The armature, therefore, soon acquires a very rapid rate of motion, which is kept up as long as the current flows. A very feeble current suffices to set the model in motion, one cell of a voltaic battery being sufficient. The velocity ultimately attained chiefly depends upon the softness of the iron in $s\ n$.

It is of little consequence which of the magnets is allowed to rotate, but in the case of a permanent magnet being used, to set up the magnetic field, it is always necessarily more bulky than the armature, and it is therefore preferable to cause the lighter body to revolve.

Attraction only.—If the field magnet were of soft iron, excited by the current, the armature might be of soft iron simply. In this case it would rotate under the influence of attraction only, and it would not be necessary to reverse the current. The current would be interrupted to allow the armature to continue its rotation past the poles under the influence of inertia simply. This is a very common and inefficient form of motor.

Electro-Magnetic Field.—When the permanent magnet is replaced by an electro-magnet, a much

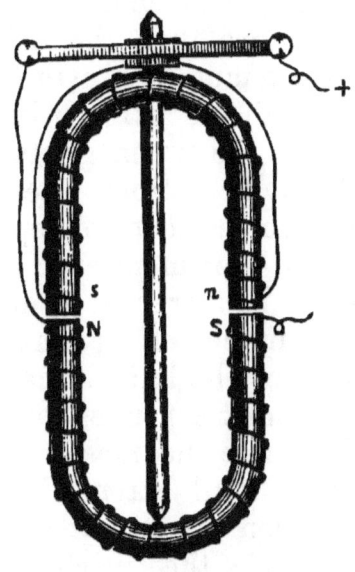

FIG. 9.

greater force is given by the machine. The apparatus then acts like a dynamo-electric machine. The magnetism of the field magnet need not in this case be reversed. The current is first passed through the field magnet, thence to the + commutator spring, and so through the armature back to the source. The action is precisely the same as before, but the rate of rotation and power is greatly increased. A short break of circuit takes place at each reversal of the current in the armature, but this does not materially effect the field magnet. More effective forms of commutator, to be described subsequently, effect the reversal without opening the circuit. The principle of such a machine is exhibited in Fig. 9, where two electro-magnets are used, one fixed and the other capable of rotation, or one as armature, the other as field magnet. It is more convenient in all such cases to rotate the armature than the field magnet, on account of the commutating arrangement, which usually rotates with the axis. Care is taken to pass the current in the two magnets, so that, when beginning to mutually attract, the poles shall be $_N^s{}_S^n$, and at the commencement of repulsion, $_N^n{}_S^s$

Froment's Model.—As mentioned above, a simple iron armature may be attracted by an electro-magnet, and the current cut off just as the former is passing the poles of the latter, the motion being maintained throughout a portion of the revolution by inertia. This was the most common form of the electro-motor until extended knowledge was brought to bear upon the machine. It is also the most inefficient form the machine can assume, because only a small percentage of the current can by its aid be transmuted into useful effect. But the electro-magnets may be two in number, and one may be active while the other is necessarily idle. Thus a *continuous attraction* may be kept up.

In its most simple form this is easily accomplished in the form of motor devised some years ago by M. Froment, and used by him for driving his philosophical instrument making apparatus. Fig. 10 exhibits the construction of such a machine, when the number of electro-magnets is two, and the number of armatures seven. The electro-magnets are ordinary U magnets, and the armatures pieces of soft iron, attached at regular distances apart around a drum capable of rotation. There being an odd number of armatures, the machine is so arranged that an armature is always on the point of being attracted by one or the other of the magnets. The electro-magnets are alternately made and unmade, as an armature approaches and then passes the poles, by a commutator on the

FIG. 10.

axis of rotation. At best this form of motor is not effective, when compared with the more recent machines, and its construction need not therefore be further detailed.

Siemens' Armature.—The invention of this beautiful device is due to Messrs. Siemens and Halske. It was introduced originally to replace the more defective forms of armature in magneto-electric machines. It may be assumed that as this armature has yielded more favourable results in generating electric currents than any preceding invention of the kind, it is admirably adapted for use as an electro-dynamic armature.

It really consists of an electro-magnet of peculiar shape. The exciting coil is wound, not

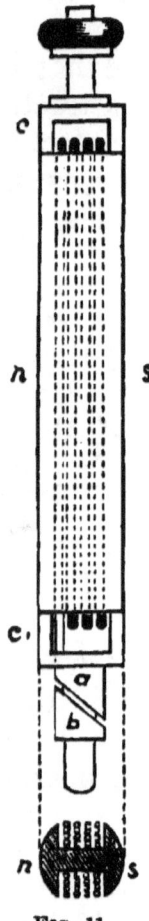

FIG. 11.

transversely, as in common electro-magnets, but *in the direction of the length* or axis. Fig. 11 will render the form of this armature more clear to the reader. It consists of a long straight iron bar.

Its cross-section calls for further explanation, however. It resembles a piece of round iron bar, *grooved out longitudinally at opposite sides.* This groove contains the exciting coil. In the section at the foot of Fig. 11 the real form is shown. The exciting coil is wound in the long groove and around over the ends of the bar in one continuous direction. Thus four or more layers of wire may be coiled in position. The poles of the magnet are not, as in other forms, at its extremities. They are situated at its rounded sides, as shown at $n\ s$. Thus it may be assumed that the armature is really an electro-magnet, but very short in the direction of its polar axis. Hence it may also be assumed that its great axial length fully compensates for its extreme shortness in the direction of its polar axis; c and c_1 are portions of the casting or body of the armature, for the purpose of receiving the axis of rotation and providing a bridge through which the exciting coil may be wound upon the " web " or connecting body of the armature; a and b show the form of commutator generally used. In effect it merely consists of two half rings, as in the form already described, but it offers the advantage of never wholly interrupting the current. It makes contact with one half before it breaks contact with the other half. This is effected by giving an oblique form to the rings, instead of dividing them direct in the line of the axis. The result of making the springs bear upon the commutator is to change the direction of the current gradually. This has another object. In the older form, when the circuit was suddenly opened a large spark was produced at the point of rupture. The result was to burn away a portion of the spring or commutating cylinder at each break. The sparking is to a great extent avoided in the above form of commutator.

The rounded sides of the armature are turned after the journals have been fixed, and the space left after coiling on the wire is filled up with packing pieces (lagging) of wood, secured by two or more encircling brass rings. In commencing to wind the exciting coil, its first extremity may be connected direct to the metalwork of the armature. It is then coiled in the groove, with precautions to insure good insulation from the body of the armature. To aid in this, the groove should be free from sharp projections, and may be coated with Japan varnish. The finishing extremity of the exciting coil is carefully insulated from the armature and attached to section *a* of the commutator, which is also insulated, by being secured on a sleeve of ebonite or wood, rotating with the axis. The second portion of the commutator is secured in metallic connection with the armature by being driven directly upon the journal, and is therefore in conductive connection with the commencing end of the enveloping coil. The opposite extremity of the armature carries the driving pulley, or pinion, when the speed of rotation is required to be reduced. The whole forms a compact armature, ready for mounting between the poles of an electro or permanent magnet. It is usually found most convenient to form the body of the armature of cast iron, which may afterwards be annealed. The armature may be of any size, according to the power of current it is intended to transform into mechanical effect. For a motor intended to produce about two-horse power, the armature should be about 36in. in length. The diameter is partly regulated by the length. An armature 36in. long may with advantage be 6in. or 7in. in diameter. In cases where the armature is of soft iron, gun-metal end plates, as in Wylde's machine, may be attached, for receiving the journals, instead of cast projections.

In forming the machine the main object is to place the armature in an intense magnetic field,

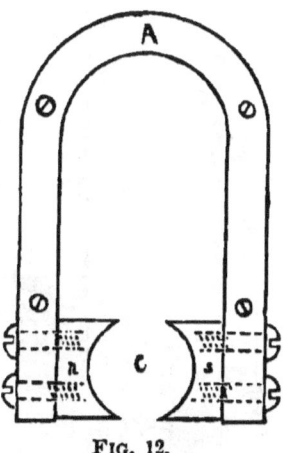

Fig. 12.

which may influence it from end to end. This may be accomplished by forming a pair of long cast-iron polar pieces or inductors, and so arranging them that a chamber may be formed, in which the armature can freely rotate. The inductors may be magnetised powerfully by attaching to them a number of permanent magnets, forming a magnetic battery.

Fig. 12 represents the end of an arrangement of this kind, where A is a permanent magnet, forming one of a series, fixed in the same position; n and s are polar pieces of cast iron, kept in a magnetised condition by contact with a, and setting up a magnetic field in the armature chamber c. The diameter of this chamber is only a little greater than that of the armature, so that the latter may rotate without actual contact with the sides of the chamber. The chamber being as long as the armature, a considerable number of permanent

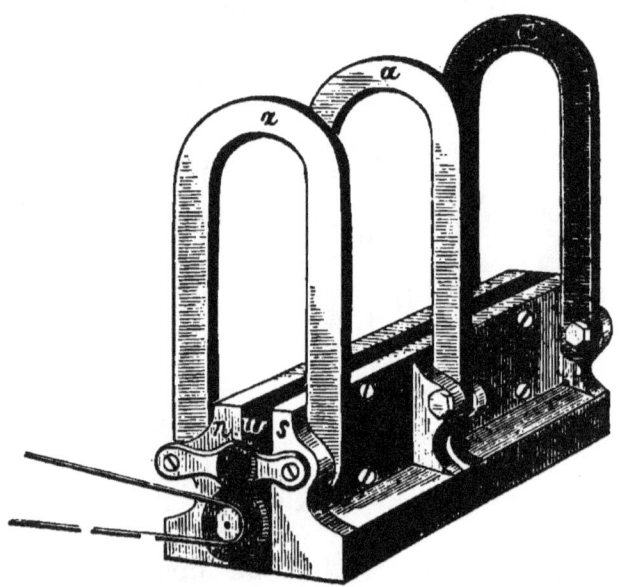

FIG. 13.

magnets are required to preserve a uniform magnetic field throughout its length. As many as sixteen or twenty magnets may thus be employed for an armature chamber 24in. in length. When only a few are used, the magnetic field is comparatively feeble, and the energy of the motive machine is but small.

Fig. 13 represents a small electro-motor arranged on this principle. The length of the armature in the original is only 14in., and only three permanent magnets are used to set up the magnetic field, but the model illustrates the principle, and furnishes a motive force of about 1,000 foot-pounds. *a a a* are the magnetic steel U's attached to a pair of cast-iron inductors *n s*, forming between them an armature chamber. The two sides of the chamber are prevented from approaching each other by a packing of wood *w*, running from one end to the other, and the whole is solidly fixed together by screws or bolts. The armature chamber is bored out "true," in the manner of finishing the interior of a steam cylinder. Packing pieces of brass may then be necessary, and it is advisable in such cases to retain them in place of the wood packing spoken of. It is needless to state that in no case can iron be employed for this purpose, since it would act in closing the magnetic circuit and annulling the magnetic field. The inductors also carry bearing pieces of gun metal for the armature, and the insulated contact springs bearing upon the commutator. At the driving end the armature carries a pinion, gearing into a larger toothed wheel, carrying concentric a strap-pulley, and reducing the velocity of rotation to one-fifth. The velocity, when a strong current, as from four or five voltaic cells, is passed into the armature, is about 3,000 revolutions per minute. The armature is wound with four layers of insulated wire, No. 16 size, and has a diameter of $1\frac{3}{4}$in.

In small models or motors, intended to yield about 1,000 foot-pounds of energy and under, the armature may be placed in a magnetic field developed by even more simple arrangements of permanent magnets. Fig. 14 represents a battery of magnets as employed by M. Deprez in his motors. A circular chamber is cut out, describing an arc upon either series of the poles, and the armature rotates

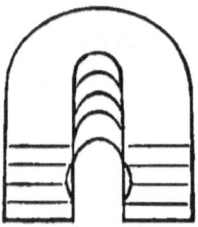

Fig. 14.

between them direct. But in this case it appears impossible to maintain the field uniform at every part of a long Siemens armature.

Fig. 15 represents a better arrangement, first employed by the author, in which two magnetic batteries are used to set up the field, and a Siemens

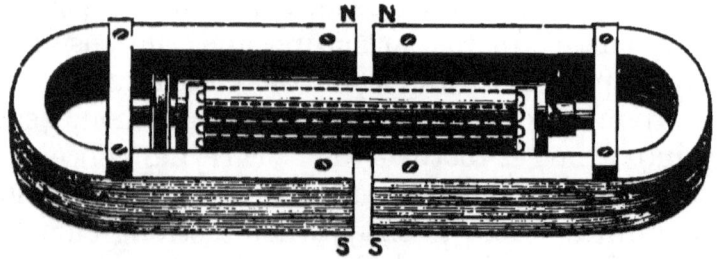

Fig. 15.

armature of the kind already described is fixed as shown, so as to freely rotate between the poles. In this case the arc shown in the preceding figure may be cut from the magnets direct, or cast-iron inductors, properly shaped to embrace each one-third of the armature, may be attached to the magnets. The axis of rotation revolves in two gun-metal bearings, attached across the magnets. Like poles of the magnets are placed together, as indicated in the figure. Magnetic batteries of this description may be made and charged by the methods detailed in previous sections.

When the electro-motive machine is intended to exert any considerable amount of energy, it is ad-

FIG. 16.

visable to replace the permanent magnets by electro-magnets. A considerable increase of power is yielded by motors when furnished with electro-magnets in place of permanent magnets. Moreover, the size and weight of the motor may be greatly diminished. The cost is much less, and the machine is capable of converting a much larger power of current into mechanical effect.

Fig. 16 represents a motor similar to the machines

already described, but furnished with an electro-magnetic field. *c* is an electro-magnet of peculiar form, the polar extremities of which are secured to a pair of inductors *b b*, as already described. The rate of rotation is reduced to one-fifth, and the power is taken off by the cord *f*.

Fig. 17 represents the same motor in cross section, and exhibits the armature, its wire envelope and packing pieces of wood occupying the field chamber.

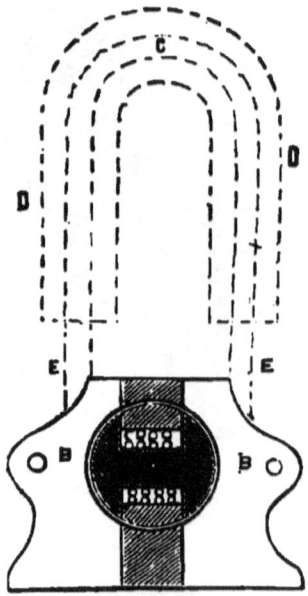

FIG. 17.

In the original, represented by the engraving, the field-magnet is composed of a piece of boiler-plate, ⅜in. in thickness, curved to the form shown. The extremities of its limbs E and E are planed, and they make perfect contact with the exterior of the inductors, which are planed also. It is very important to insure that all "magnetic connections" shall be perfect. If the surfaces are rough or oxidised, the magnetic continuity will be broken, and the magnetic field weakened. The field magnet in question is excited by a coil of wire enveloping almost every part of it, and four layers deep. It is composed of No. 14 insulated wire. The current leads, first, through this coil and then to the first contact spring, from which it passes through the enveloping coil of the armature, and back to the electric source by the second contact spring. The motor is furnished with an armature 14in. in length. With a current of about five webers, the effective motive power is about 1,300 foot-pounds. The electrical resistance of the armature coil is 0·4 ohm, and that of the magnet coil 1 ohm, a total of 1·4 ohm, which is, however, considerably increased when the motor is in action.

Starting from the position shown in the figure the armature is repelled by the magnetic field through one-fourth of its revolution. This brings the poles within the influence of the reverse polarities of the field, and the armature is attracted through another quarter of a revolution. The position is now again as represented in the figure, but the poles are changed. At the instant of reaching the medial line of the magnetic field, the current is reversed, and repulsion ensues, then attraction, then another reversal, and so on, as in the cases previously described. Fig. 18 exhibits the same form of motor, furnished with a permanent magnetic field.

The extraordinary efficiency of this form of motor is very greatly due to the fact that the armature has relatively very little motion. No fraction of an instant is allowed to pass without throwing duty upon the magnetic field. Thus, in the model represented at Fig. 7 there are considerable portions of the revolution, or a considerable percentage of the total time, during which the actuating current is allowed to remain idle. The Siemens form of armature fully utilises the power of the current during each revolution. Its relatively small motion is compensated for by its peculiar prolonged form, which makes it equal, in effect, to a great number of small magnets rotating upon one axis.

Fig. 18.

The engine represented in the upper portion of the first page is designed in a manner similar to that already described. It is furnished with an electro-magnetic field of boiler-plate, connected to two inductors *b b* as before. The contact edges and faces are planed to insure continuity of magnetic circuit. The armature is of the kind already described. The current is reversed by the commutator at each half revolution, by means of the adjustable contact springs. The current enters from the source at +, passes around the field magnet, and so to the first contact spring. Thence it flows, by way of the commutator, through the armature-exciting coil, and so, by way of the second spring, to the electric source. *g* is the strap by which the power is taken off for external purposes. The base *a a* must not be of iron, unless separating pieces of brass be employed, with brass bolts, in order to break the magnetic continuity between the inductors.

Fig. 19 represents this motor in section *c* is the

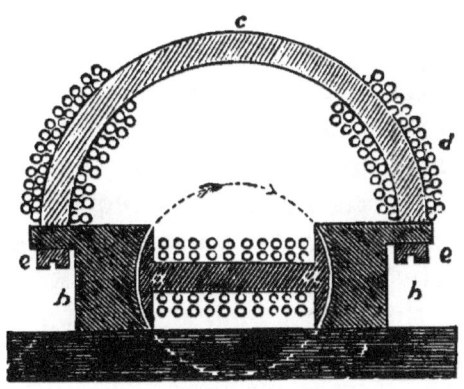

FIG. 19.

core of the field magnet, the exciting coil of which is divided into two sections, wound upon each side of the core; *d* the wires, two layers being represented, but in the machine itself there are four layers; *a a* is the armature, with its wire coil of four layers (two layers only are shown); *e e* are screws or bolts making the magnet core fast to the inductors.

The engraving in the lower portion of the opening page of the work is of M. Trouve's electro-motor, employed by him for the propulsion of light pleasure boats, tricycles, etc. It is also furnished with an electro-magnetic field, which is created between the inductors *a a*. The armature revolves in this chamber. *b* is a portion of the electro-magnet, which is excited by the wire coil *c*, wound around its rectangular or U bend. The current enters at +, thence passes by the commutator around the armature coil, and from this by the lower commutator spring to the field magnet, after exciting which it flows back to the electric source by the terminal marked —. The useful motive power of the machine is taken off from a pulley by the band. When the pulley is small, this band is crossed to afford a better bearing upon the circumference.

In these machines, when large, a bundle or spring of hard-drawn copper wires will usually be found to make a more satisfactory commutator spring than a single piece of brass or copper. Copper suffers less by sparks than brass or steel. Platinum suffers less than most metals, but it is usually too soft to stand the necessary friction. Iridium, which is not much affected by the sparks, and is exceedingly hard, would probably answer best, but it is at present too costly, and is not readily obtainable in the required form. Hard-rolled thin sheet copper may be used in some cases, in numerous *laminæ*, arranged so that the bearing ends shall form an angle against the commutator.

The brushes must in all cases be so set as to reverse the current at the instant when the armature poles are passing through the medial line of the magnetic field. When the velocity of rotation is very great, it will be found that the theoretical line of reversal must be departed from. The brushes must then be adjusted so as to compensate for the time necessary to magnetise and demagnetise the armature. In most cases the reversal of the current must be effected in advance of the theoretical point. When it can be effected, both contact brushes should be fixed in a bracket with two branches, capable of motion on the axis. By these means any movement of one brush is communicated to the other, and the brushes are maintained at diametrically opposite points of the commutator.

In the construction of the machines, insulation, where insulation is necessary, must be *complete* and unmistakable. This refers to magnetic as well as electrical insulation. The insulating covering of the wires may be silk or cotton for small machines, preferably after being run through melted solid paraffin. The sizes of wire for small machines vary from No. 12 to No. 18. For the larger machines the wires may be covered with cotton or hemp, and treated either with varnish or paraffin. Shellac varnish is however preferable in cases where heat is likely to be involved. The sizes of wire may vary from No. 6 to No. 16. In most cases where wire is wound upon iron direct, the surface should first be freed from roughnesses likely to chafe the insulating covering, and the angles should be rounded. The surface should also be treated with a coating of Japan varnish, baked on, or applied while the iron is hot. Guttapercha, wood, and ebonite should be used to insure insulation. Ebonite should not be used in the construction of commutator cylinders, because the heat developed by friction softens it. When a wire is attached to another wire, or to the commutator, and in every case of effecting circuit junctions, care must be taken to insure that the connection is metallic and clean. The surfaces must also be pressed closely together, otherwise oxidation may be set up and the junction destroyed. In many cases the ends of wires should not only be twisted together, but soldered also. The resistance offered by bad joints may reduce the efficiency of the machine enormously.

Magnetic insulation is more easily effected than electrical insulation—that is, it is only necessary to use either wood or other non-metallic substance, or those metals not affected by magnetism, such as brass and copper. Magnetic continuity is less easily effected. When two surfaces come in contact, and when it is desired to pass magnetic polarity through the junction, as in the case of field magnets and polar inductors, the surfaces should be planed or otherwise rendered level to insure their touching all over the area. The parts must also be firmly secured together by mechanical means, iron screws and bolts being used in preference to copper or brass.

In the case of the Siemens machines employed in the propulsion of railway carriages, the velocity is reduced to one-third, by means of a chain connection. In most cases, where these and other machines are used it is advisable to reduce the velocity; but this may be done in many instances without any reducing gear on the machine itself. The velocity of the armature is necessarily very great, and care should be taken to provide means by which the maximum speed may be freely developed, otherwise the effective power of the machine may be reduced.

Siemens' Modified Armature.—Enough has already been said to show why the ordinary Siemens armature yields so large a return of power for the current expended in moving it, and why it must be regarded as an electro-magnet of peculiar shape. At each revolution of the armature there are two attractions or periods of attraction by the force in the magnetic field, caused by the flow of direct currents around the armature; and two repulsions or periods of repulsion by the force in the field, caused by the flow of inverse currents. Extending the idea of the assumption of polarity by the poles of the armature alternately in one given direction, it is not difficult to believe that if a *complete cylinder of iron* were used instead of a rail-shaped bar, each portion of it, when rotated in the magnetic field would *successively* become either N or S; or an N and S polarity would be constantly maintained, whether the armature moved or remained fixed. These induced polarities would of course be opposed to the polarities of the magnetic field, since, when such a cylinder of iron is rotated in the field each portion of it becomes magnetised, it is clear that if the cylinder were completely enveloped with a layer of insulated wire, wound longitudinally, currents corresponding to the assumption of polarity by the underlying iron would be developed in the wire— that is, when the armature is used as a generator. When employed as a motor, the attractions and repulsions of the simpler armature would ensue, but in this case, since the iron and enveloping coil are continuous, it may be assumed that a *constant unbroken* strain of attraction and repulsion would be kept up between the magnetic field and the armature. This is, in fact, the Siemens modified armature, now used in the dynamo-electric and the electro-dynamic machines of that type. It consists of a complete hollow cylinder of iron, long and narrow, like the simpler armature, and rotating on an axis running in the direction of its length. The surface of the armature is completely enveloped by insulated wire, wound in sections over the sides and ends, and communicating with a commutator on the axis. On account of the inductive influence upon the iron cylinder being exerted at diametrically opposite points of its surface, the two sides of

FIG. 20.

the cylinder between these points may be assumed to form the body or central portion of an induced magnet, so that there must ensue a mutual interchange of force between the two opposite points occupying at any moment the medial line of the magnetic field. Hence, the sections of the wire enveloped are connected together in a certain manner, which may be understood by reference to Fig. 20. N and S represent the poles of the field magnet, placed in this case in a vertical line passing through the axis of the armature. The central portion of the diagram represents the end of the iron cylindrical armature, and the arrangement of the commencing and terminating extremities of each of the wire sections spoken of above *c c*, etc., are a series of twelve plates, usually of copper, arranged parallel with the axis of an insulating commutating cylinder. Each plate is furnished with two connecting screws to receive the wires as shown. The wire sections are so connected to the plates that, while the commencing end of No. 1 section is fixed to No. 1 commutator plate, the finishing end of the same section is carried round to the opposite plate as represented. This is carried out all round the circle. Collecting brushes press upon opposite diameters of the cylinder, and supply the current (or collect it when the machine is used as a generator). The field magnet is composed of a series of iron bars, curved to the form represented in Fig. 20, and excited by four coils. The current first passes through the field magnet, and from it to one of the commutator brushes, thence through the armature and back to the electric source by the opposite commutator brush. Other details of the construction of this machine are so amply treated upon in treatises on electric lighting, in which also drawings are provided, that a fuller description of it here would be deemed superfluous.

Gramme's Armature.—This armature consists of a ring of soft iron, capable of rotation in a vertical plane. Its surface is enveloped by an endless helix of insulated wire, several layers deep. Fig. 21 will render this arrangement more clear, where the ring is represented as free to rotate in a magnetic field induced by the electro-magnet. The connection between the armature helix and the external circuit in this form of machine is different from that in Siemens's machine. As already mentioned, the ring helix is complete. It practically forms a closed circuit wound upon every part of the ring. Each turn of wire represented in the diagram stands for a section of the complete helix. The connections to the commutator are as numerous as the coils. One connecting wire leads from the commencement

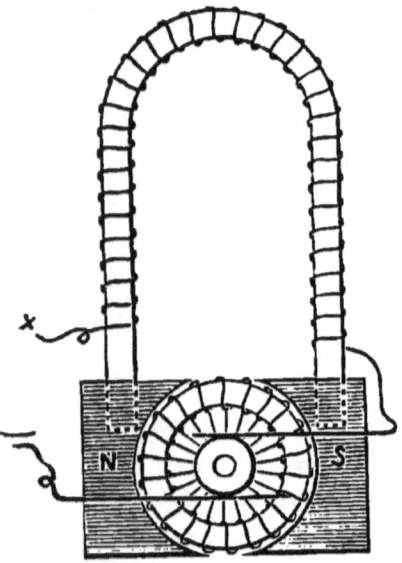

FIG. 21.

of each section to a corresponding contact plate on the commutator cylinder, as in Siemens' machine. Each finishing end of each section leads, however, in this case, to the beginning of each succeeding section. Hence, the circuit between the commutator brushes consists of all the helical sections on one half of the ring, or on either half, or on both halves together, forming a divided circuit, according to the direction of the current.

As the ring is caused to rotate (employing the machine as a generator) in the magnetic field, opposite polarities are induced in diametrically opposed portions of the ring, as in the modified Siemens armature. Each of these diametrically opposed portions may therefore be regarded as an electro-magnet, in which a current is induced. These currents are assumed to flow towards each other and to meet at points equidistant from the poles of the magnetic field. They are assumed to act as in a divided circuit, flowing through the two halves of the ring, meeting and forming a uniform current in one direction at the commutator brushes. If the commutator brushes were removed, no useful currents would be developed, since they would be neutralised by meeting opposite currents. The portion of the circuit between the brushes (the external portion) therefore completes the circuit necessary for the development of currents by the ring.

It may be assumed that as the Gramme armature, when rotated in a magnetic field, gives rise to a continuous flow of current in one direction, any current passed into it instead will produce a constant attraction and repulsion between the magnetic field and the magnetism of the ring. This attractive force will be so exerted that motion will ensue, and the repulsive force will conduce to the continuance of this motion. Therefore the Gramme armature may be regarded as eminently adapted for use as an electro-motor. It may be assumed that at no part of the revolution is the tendency to motion more or less powerful than at any other part of the revolution.

In the Gramme machines the armature is composed of a ring core of soft iron wires, forming, however, a complete circle. The wire is wound on in sections several layers deep. A great number of these sections are employed. The commencing extremity of each section is connected to a copper commutator plate. These plates are arranged radially around the axis, towards one side of the ring, and their edges project slightly from the cylindrical surface of the commutator insulating cylinder. 1,000 foot-pounds, or the thirty-third part of a horse power is a common motive power for a Gramme motor of small size (5in. armature). The ordinary Gramme light machines, of the small size, requiring, when used as generators, an expenditure of about three horse-power, yield as motors, through a moderate resistance, nearly $1\frac{1}{2}$ horse power.

Reciprocating Motors.—These are in most cases very inefficient. Attraction to the poles of a magnet in a direct line, even if followed by repulsion, is wasteful of energy, inasmuch as the distance moved through is exceedingly small, and a very considerable number of strokes must be taken by the armature in a short time to produce much effect. The principle of the solenoid is better; but the interactions which take place when a plunger is sucked into a hollow coil and then expelled are almost identical with those involved in the motion of the Gramme armature. Since, therefore, the principle of the action in reciprocating motors with solenoids can be utilised by the rotatory movements of Gramme's armature, it is evident that the reciprocating motor is under great disadvantage, reversals of direction of motion being wasteful of energy. In fact, one revolution of a Gramme armature may be assumed to be equal to a great many strokes of a reciprocating engine of this kind, actuated by an equal current. In any motor where the moving armature is attracted simply to an electro-magnet, and then allowed to continue its movement by demagnetising the magnet, and inertia, a great loss must occur. The residual magnetism tends materially to retard the progressive movement of the armature.

The Inverse Electro-motive Force.—It is generally assumed that all efficient electro-motors tend to develope an electro-motive force reverse of the driving current. This inverse electro-motive force may be measured as resistance, so that as the inverse force increases the direct current decreases. The result is assumed to be increased efficiency. Differences of opinion, however, exist as to the real effect of the inverse force, some authorities maintaining that it must necessarily lead to a loss of energy, and others holding that it is this back electro-motive force that makes the machine a motor at all.

On this latter point, Prof. Ayrton, F.R.S., says, " This induced current (inverse force) therefore must be taken into account in calculating the efficiency of an electro-motor—in fact, not only must be taken into account, but experiment shows us that it is this tendency of the electro-motor to send a back current that makes it an electro-motor at all. . . . When an electro-motor is worked by a *given* galvanic battery, calculations lead us to the result that if we wish to produce the work most economically we must, by diminishing the load on the motor allow, its speed to increase until the reverse current it produces is only a little smaller than that sent by the battery—in fact, until the current circulating through the arrangement is very small, in which case the efficiency of the engine, or the ratio of the work it produces in a given time to the maximum of work it could produce from the same consumption of material, is nearly unity. If, on the other hand, we desire a *given* battery to cause the motor to do work most quickly, independently of the consumption of material, then calculation tells us that we ought to put such a load on the motor that its speed will send a reverse current equal to something like one-half of the strength of the current the battery could send through the motor when at rest. In this case the efficiency is about one-half, or half the energy is wasted in heat." (British Association Meeting at Sheffield, August, 1879.)

Dr. Werner Siemens, on the same point, speaks as follows: " When one dynamo machine is driven by another, the driven will generate a current in the opposite direction to that produced by the driving machine, from which consideration it is evident that the maximum power must be produced at a certain speed for the driven machine; and by the author's calculation for two *similar perfect* machines this would be when the driven machine is running at one-third the speed of the other. By a perfect dynamo machine is understood one in which the mass of the iron is such that the magnetic intensity is proportional to the current in the convolutions, and in which, when the current is closed in itself, the work is proportional to the third power of the velocity of rotation. In practice this is not so, owing to the increase in the resistance of the brush contacts, and the molecular resistance of the iron to magnetisation at the higher velocities. This being the velocity at which the maximum work is produced does not mean the efficiency of the machines is only one-third, but the power reproduced in useful work is proportional to the velocities of the two machines. From numbers of experiments recently made at moderate velocities, the useful work was 40 to 50 per cent. of the expended, and at higher velocities as much as 60 per cent. was obtained. It is perhaps merely a question of the size and velocities of the driving and driven machines whether a still higher percentage of power may not be regained."

On the same subject, and dealing specially with the question of the relative speeds of the driving and driven machines, Dr. C. W. Siemens, F.R.S., at a special meeting of the Society of Telegraph Engineers and Electricians, said that "It is a remarkable circumstance in favour of the electric transmission of power that while the motion of the electro-magnetic or power-receiving machine is small, its potential of force is at a maximum, and it is owing to this favourable circumstance that the electric train (in the case of the electric railway at Berlin) starts with a remarkable degree of energy. With the increase of motion the accelerating power diminishes until it comes to zero, when the velocity of the magneto or driven machine becomes equal to that of the dynamo or current-producing machine. Between the two limits of rest and maximum velocity, the driving power regulates itself according to the velocity of the train; thus on an ascending gradient the speed of the train diminishes, but the same effect is automatically produced which results from the turning on of more steam in the case of the locomotive engine. When running on the level, the velocity of the train should be such that the magneto-electric machine should make one-half to two-thirds as many revolutions per minute as the dynamo-electric. When descending, the speed of the magneto-electric machine will be increased, in consequence of the increased velocity of the train, until it exceeds that of the dynamo-electric machine, from which motion the functions of the two machines will be reversed: the machine on the train will become a current generator, and pay back, as it were, its spare power in store, performing at the same time the useful function of a brake in checking further increase in the velocity of the train. If two trains should be placed upon the same pair of rails, the one moving upon an ascending portion, the other upon a descending portion of the same, power would be transmitted through the rails from the latter to the former, and the two might therefore be considered as connected by means of an invisible rope. . . . In transmitting the power of a stationary engine to a running train, the proportion of power actually transmitted varies with the resistance or speed of the train, reaching practically a maximum when the velocity of the machine on the train is about equal to two-thirds that of the current-generating machine, at which time more than 50 per cent. of the power of the stationary engine is actually utilised."

CHAPTER VIII.

Electric Railways.

The development of the principle of converting motive energy into powerful electric currents, led to the idea of transmitting power through conductors to any required distance. This could easily be accomplished by the aid of the correlative principle of transforming electric currents into motive power. It is evident, for example, that if a dynamo-electric machine produces an electric current of the power of ten horses, this current may be transmitted through a suitable conductor to a certain distance, dependent upon the resistance of the conductor, then passed through an electro-motor, so as to reproduce the energy as motive effect. This cannot be done without a certain loss. But neither can the steam engine be worked without enormous loss. The loss sustained in the case of electric transmission of power is very small when compared with that involved in the transmission of energy by means of compressed air, water, or other ponderable fluid. The efficiency of the new means suggested by the production of powerful electric currents at small cost having been found to be higher than that of compressed air or other method, led naturally to the conception of an electric railway, or a system of conveyance in which the carriages would be propelled by an electro-motor. One fact was of supreme importance in this connection. It was that the electric energy might be transmitted along the line through either of the rails, and that it might actually be supplied to a moving electro-motor through a *frictional contact*. Such a method of communicating energy would be impossible by any other means known to man. Thus it was conceived that it was unnecessary to convey in the carriage itself the materials for producing the electricity, or even the electricity itself in a condensed or stored form. The actual energy could be produced by a stationary machine, situated at any distance from the train itself. It was only necessary to allow a brush of metal, or other suitable collector attached to the train, to continuously touch the current-bearing rail. By these means the electric energy could be passed through the electro-motor, so completing the circuit by the other rail. This immense advantage included many others. It offered to allow of a considerable diminution of weight in the rolling-stock of railways; it shadowed forth the possibility of doing away with the ponderous steam locomotive, carrying its own energy; it promised to afford a means, hitherto impracticable, of immensely increasing the tractive power of a railway train, so that it might in safety ascend steep gradients and start rapidly. This could clearly be accomplished by distributing the motive force along the train. In short, by entirely abandoning the use of a separate locomotive, and fitting an electro-motor to each carriage, which would thus be capable of moving itself. With reference to this point, Captain Galton, years before electric railways were thought of, suggested, in one of his official reports on the brake power and high speed of trains, that

"The advantage which thus evidently ensues from utilising the adhesion of every wheel of a train (in brakes) suggests the further consideration as to whether it would not be a more scientific arrangement, as well as more economical in regard to the permanent way of railways, to utilise the adhesion of every wheel of a train for causing a train to move forward, instead of depending for the moving force upon the adhesion of one heavy vehicle alone, viz., the locomotive."

The advantage of thus attaching an electromotor to each axle of the train, or at least to each vehicle, would result in greatly-increased safety to the passengers. It would result in less damage to goods, less wear and tear generally, and add to the facility of passing safely round curves. The electric railway also promised to afford almost absolute immunity from accidents, by means of the facility with which the current upon each section of the line might be instantly suspended in case of danger. It might also be instantly reversed, so causing two approaching trains to come to a stand and move in the reverse direction. It will be evident from this consideration that no brakes would be necessary or so effective, in the case of an impending collision, as the instantaneous reversal of the motors throughout the train. The lightness of the train would greatly contribute to a rapid stoppage, evenly distributed. In addition to these more substantial advantages, the proposal to propel trains by electricity derived from the fixed rails promised to do away with the nuisances attending the use of steam locomotives. Since no energy would be created in the train itself, there would be no waste, such as ash, smoke, and steam, to dispose of. Thus the contamination of the atmosphere, more especially with reference to underground lines, could be entirely avoided, and the deterioration of the ironwork of tunnels and bridges by the gaseous exhalations of the steam locomotive would entirely vanish.

Some authorities have given it as their opinion that trains might be propelled by electricity as cheaply as by means of the steam locomotive. They base their argument, first, upon the great expenditure of coal per horse power in the working of steam locomotives, an expenditure from five to ten times as great as that incurred per horse power in the working of large compound condensing stationary steam engines; and, secondly, upon the high efficiency of the dynamo-electric machine, which converts over 90 per cent. of the power of the steam engine into effective electromotive currents. We have seen from the experiments carried out to determine the efficiency of the dynamo-electric machine when used as a motor, that at present the energy of the motor is nearly 50 per cent. of that expended in driving the current-generating machine. By these means it would appear to be possible to actually propel railway trains by electricity at an expenditure for power, through moderate resistances, little if any greater than that at present incurred by the use of steam locomotives. The element of resistance is, however, one of great importance. To secure results approximating to those proposed, the leading rail or conductor would require to be of small resistance, and its insulation effective for currents of low tension.

The idea of an electric railway was first mooted

by Dr. Werner Siemens, at Paris, in 1867, when he discussed with the members of the Paris Exhibition jury the possibility of elevated railroads, worked by the electrical method of transmitting power. It is only about five years ago, however, that opportunites presented themselves to Dr. Siemens to actually develope his ideas upon this subject. The starting-point appears to have been a request made by the owners of a coal-mine to be supplied with an electric locomotive for conveying coals in the mine. The result of this first practical experience of the propelling power of electricity was that Dr. Siemens constructed a small electric railway, and exhibited it on the occasion of a local exhibition at Berlin, in the summer of 1879. This railway was subsequently exhibited at Düsseldorf, Brussels, and in London (at the Crystal Palace in the summer of 1881).

This railway, as arranged at Berlin, was circular. It had a total length of about 350 yards, and the gauge was about 3ft. 3in. (1 metre). The two rails being laid upon wooden sleepers in the ordinary manner, were sufficiently insulated to serve as conductors of the current, but it was considered more satisfactory to fix a bar of iron on wooden supports, centrally between the two rails, and running parallel with them. From this bar the electric current might be taken off by the train by means of a friction brush, or a metal pulley bearing upon the conductor.

On the grounds near by was fixed a dynamo-electric machine, driven by a steam engine, the expenditure of energy being about five horse power. The current from this fixed machine was led to the conducting rail. Upon this railway a train of four carriages was placed. Upon the first carriage was mounted a medium-sized dynamo-electric machine, capable of exerting a motive power of three horses. The revolving axis of the machine was connected to the axles of the carriage, so as to impart its motion to them. A metallic brush fixed to the electric locomotive collected the current from the conducting rail, and led it through the motor in the usual manner. Between twenty and thirty persons could be accommodated in the train, the conductor riding in the first carriage, to which the form of a small locomotive was given. The electric locomotive having been fitted with a commutator, by means of which the starting, stopping, and reversing might be effected, the train was under perfect control of the conductor. When the carriages were prevented from moving, the motor exerted a pull upon them of about 2cwt. (200 kilos.), and when the train was in regular motion, the pull varied between $1\frac{1}{2}$cwt. and $1\frac{3}{4}$cwt. (70-80 kil.), which represents, at the normal speed of 10ft. per second, nearly three horse power.

A velocity of between fifteen and twenty miles per hour could be easily attained on this railway. Crowded trains left the station every five or ten minutes, and the pennies paid for the privilege of a seat in the train produced a considerable sum for the benefit of local charities. The railway was in constant operation in Berlin for several months, when it was removed to Brussels.

Regarding the working of this railway, Dr. Werner Siemens remarks: "Here it was noticed how slight a variation of speed was caused if the cars were doubly or trebly loaded or empty." Dr. C. W. Siemens recently said, "It is a remarkable circumstance in favour of the electric transmission of power that while the motion of the electro-magnet or power-receiving machine is small its potential of force is at its maximum, and it is owing to this favourable circumstance that the electric train starts with a remarkable degree of energy."

With the increase of motion the accelerating power diminishes until it comes to zero, when the velocity of the magneto or driven machine becomes equal to that of the dynamo or current-producing machine. Between the two limits of rest and maximum velocity the driving power regulates itself according to the velocity of the train. Thus on an ascending gradient the speed of the train diminishes, but the same effect is automatically produced which results from the turning on of more steam in the case of the common locomotive. When running on the level, the velocity of the train should be such that the electro-motor should make one-half to two-thirds as many revolutions per minute as the dynamo-electric machine. When descending, the speed of the electro-motive machine will be increased, in consequence of the augmented velocity of the train, until it exceeds that of the dynamo-electric machine. From this motion the functions of the two machines will be reversed: the machine on the train will become a current generator, and pay back, as it were, its spare power in store, performing at the same time the useful action of a brake, in checking further increase in the velocity of the train. If two trains were placed upon the same pair of rails, the one moving upon an ascending portion, the other upon a descending portion, power would be transmitted through the rails from the latter to the former. The two trains may thus be considered as connected by means of an invisible rope. (The substance of these views of the relative action of the two machines, as enunciated by Dr. Siemens, is also given in the chapter devoted to the treatment of the efficiency of machines.)

In transmitting the power of a stationary steam engine, by means of an electrical current, to a moving train, the proportion of energy actually transmitted varies with the resistance or speed of the train. It reaches a maximum when the velocity of rotation of the machine on the train is equal to two-thirds that of the current-generating machine, under which circumstances more than 50 per cent. of the power exerted by the stationary steam engine is utilised in moving the train.

The experience gained in the working of the Berlin experimental railway, in 1879, led to the conception that the electro-motor might be utilised to greater advantage by dispensing with a special locomotive and attaching a motive machine beneath each car of the train, with the wheels of which it would be connected. Small as the electric railway was it clearly demonstrated the practicability of this method of propulsion. It illustrated the advantages of employing light carriages, and of being able to propel them without noise or smoke.

These considerations led Messrs. Siemens and Halske to lay before the authorities of Berlin a plan for an elevated electric railway through one of the streets in the city. It was to be 6¼ miles in length. It was proposed to erect along the kerbstones of the street iron columns, formed by channel irons, about 11 yards apart, and carrying wooden sleepers on the top, which in their turn supported longitudinal girders, to insure stability of the structure. Wooden struts keep the girders apart, and serve at the same time to insulate them from each other. The clear level height, from the level of the street to the underside of the girder, is about 14ft. 6in. (4·4 metres), and the depth of the girder about 6in. (40 cm.). Steel rails are laid on top of the girders, and the girder and rail on one side serve as the conductor from the primary machine, the other rail and girder forming the return portion of the circuit. By these means the electrical resistance of the line is reduced to a very low figure.

The gauge of the line was one metre, and the carriages, resembling ordinary tramcars, were to be about 5ft. 5in. broad (1·65 m.), and 8ft (2·46 m.) high above the rails. The electro-motor, placed underneath each car, imparts its motion by means of belts to the two wheels, which must be insulated from each other, since the current arrives by one rail, passes through the machine, and returns by the other rail.

The speed at which these carriages were intended to travel is 30 kilometres (18·6 miles) per hour. Ten of them were to be placed upon the railway, of which number six would be in use and four in reserve. Ten horse power was to be supplied for driving each primary machine of each

K

carriage. The cost was carefully calculated from the experience already gained, and as it will serve as an indication what such railways may be expected to cost, a short summary of the principle items will not be out of place.

COST OF 6¼ MILES (10 KILOMETRES,) ELEVATED RAILWAY, SINGLE LINE.

Railway itself, including 10 stations	£61,000
Ten carriages to accommodate 15 persons each	3,150
Stationary steam engine and dynamo-electric machine	1,950
Buildings	1,185
Land	4,500
General expenses	715
	£72,500

or about £11,600 per mile. This estimate includes the cost of erection of the railway, and of the station at which the steam engine works, together with the necessary buildings to protect the rolling-stock against the weather when not in use. The cost of working the railway was calculated to be for one year.

CURRENT EXPENSES.

Wages	£2,190
Fuel	1,110
Oil and Waste	50
Light	80
	£3,430

DEPRECIATION AND REPAIRS.

3 per cent. on £62,500 (Railway and Buildings)	£1,875
16 ,, on £5,000 (Carriages and Machinery)	800
	£2,675

INTEREST ON CAPITAL.

5 per cent. on £75,500	£3,625
Total cost per annum	£9,720

Or about £4·6 per mile per day. The intention was to run about 200 trains per day, and if the charge of 1d. per mile had been made, the £4·6 per mile could have been earned, if on the average five or six persons had been conveyed in each case.

The necessary concession for this electric railway was not granted, partly because the inhabitants strongly objected to having people looking into their first-floor windows, and partly because the Emperor did not wish to see the "Linden," which the railway had to cross, disfigured.

Subsequently, however, Messrs. Siemens and Halske obtained permission to build a railway on the ground level from Lichterfelde, a suburban station on the Berlin-Anhalt Railway, to the Military Academy in Berlin, and this railway was successfully opened for regular traffic in May, 1881.

So great has been the success attending the working of the railway that it has since been considerably extended. It was in the first instance extended to Tetlow, and afterwards as far as Potsdam. It is being further extended to Steglitz. The railway at present works regularly without a hitch.

The first section, that to Lichterfelde, is a simple line of a gauge of 1 metre, and a little over $1\frac{1}{2}$ mile in length. The permanent way has been constructed in exactly the same manner as that of ordinary railroads. Wooden sleepers and steel rails are employed, the rails being connected, in addition to the usual fish-plates, by short straps of iron, curved in the shape of a bridge, so as to admit the adjustment of the rails to different temperatures, and to diminish at the same time the electrical resistance. As the currents are low-tension currents, it is not necessary to provide further insulation, and no difficulty is experienced in using one rail as the positive and the other as the negative conductor.

With special reference to the Lichterfelde section of the line, which was first put into practical operation, about a third of a mile from the Lichterfelde station, the primary machine with its steam engine is erected in the engine-house of the Lichterfield water-works, and the current is cenveyed thence to the rails by means of two underground cables. The cars are exactly similar to the ordinary tram horse-cars, and are constructed to accommodate twenty persons besides the guard. One car is run at a time, but two or more may be placed upon the rails if required. The car is provided with gear, by means of which it can be caused to move in either direction. It is also provided with a brake and a signal bell. The electro-motive machine is fixed beneath the car, and is connected to the axles by means of spiral steel springs, by which means a certain momentum may be acquired by the motor, before the resistance of the car is thrown upon it. These means also cause the car to start without a jerk. Either belts or a chain may be used, and are being employed in other instances for the same purpose. The tires of the wheels are insulated from the axles, and are in electrical connection with brass rings fastened upon the axles but insulated from them. Contact brushes press against these brass rings and take off the current, conducting it in this way through the motor machine. The starting gear merely consists of a circuit completer. The reverser is a commutator reversing the direction of the current.

The authorities were for some time doubtful how to class this novel railway, and after long deliberation they decided to rank it as a one-horse tramcar. In consequence of this decision, the average speed of the electric car must not exceed 9·3 English miles (15 kilometres) per hour, and the greatest speed at any moment must not exceed 12·4 English miles (20 kilometres) per hour. The time for traversing the whole distance on the Lichterfelde section is therefore not less than ten minutes, although the car could make the journey in half that time in perfect safety. When necessary, the electro-dynamic machine on the cars can be short-circuited, so causing them to act as very powerful brakes.

When a special permanent way can be provided for electric cars, as in the case just spoken of, the insulating power of the wooden sleepers is found to be sufficient. Various means may also be used to increase the insulated condition of the rails, such as chairs of glass, guttapercha or felt packing, or other like method. But in the case of running a line of this kind after the manner of a common tramway in streets, the insulation is not sufficiently good. This difficulty, it has been proposed, can be overcome by the use of tall posts, placed at intervals along the line, and carrying a naked conducting cable. A contact or trolly arrangement, communicating with the car by a cable, which also serves to move the trolly forward, is proposed for use with this; but as such a system would be both objectionable on common roads and impossible in streets, it will be necessary to devise a means of taking the current from a conductor on or beneath the surface of the street. The most likely plan appears to be by means of a copper or iron conductor, which could be attached almost underneath one of the rails, and insulated from it with felt packing. If the rail were of a special shape, with its inner side cut away, forming, as it were, a protecting flange for the conductor, it might prove practicable by these means to take the current off by means of a yielding spring brush or rolling contact maker. The system might, of course, be duplicated, in the case of difficulty being encountered by leakage from one rail to the other in wet weather, and means could probably be taken to prevent any such leakage when low-tension currents are employed. At the Electricity Exhibition held in Paris in the autumn of 1881, Messrs. Siemens and Halske ran an electric tramcar regularly on the ground-line from a wooden station on the Place de la Concorde into the exhibition, which was held at the Palais de l'Industrie in the Champs Elysées. In this case the car was of the same construction as is used in the Lichterfelde line. It had an electro-motive machine attached beneath. The axis of rotation was connected with the two wheels of the car on one side by means of two chains, and the relative diameters of the axis and wheels reduced the velocity of rotation to one-third (Fig. 27). In the exhibition building the primary dynamo-electric machine was situated. It was at first tried to work the car by means of taking the current direct from the rails, but it was found that in an open street this could not conveniently or effectively be

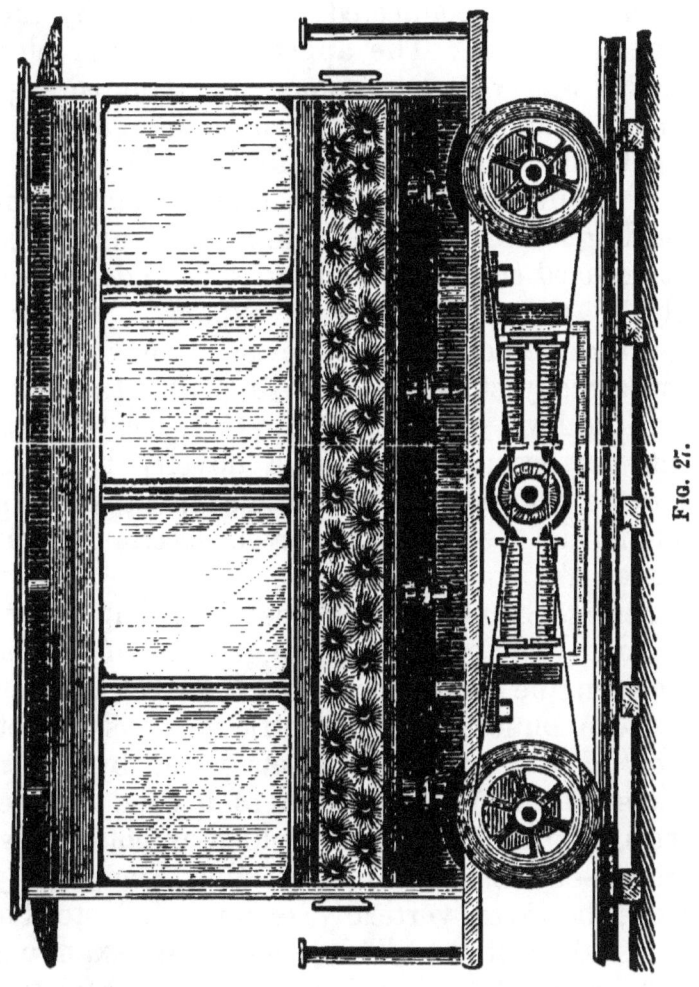

Fig. 27.

accomplished, since no special precautions were taken to effectively insulate the rails. It therefore became necessary to erect a series of strong iron pillars, carrying on the top a naked conductor, after the manner of a telegraph wire. From this the current was conveyed to the machine in the car by means of a trolly running on the elevated conductor, and connected with the car by means of a wire cord. The appearance of the electric tramcar was very handsome. It attracted a great deal of attention in Paris, although similar cars had been regularly open for traffic in Berlin many months before. The method employed for "feeding" the motor machine with current could not be used commonly, and it would add considerably to the constructional expense. It would therefore be desirable to devise a means, as we have already suggested, for supplying the current from the level of the street. It is very probable that the exhibition of this novelty by Messrs. Siemens will lead to the establishment of a permanent elevated or street-level electric railway in or near Paris.

Messrs. Siemens and Halske now employ a very compact and portable arrangement of combined steam engine and primary dynamo-electric machine for use in connection with electric railways, or in other cases of the transmission of power by electrical means (Fig. 28). It consists of a solid and heavy casting, forming a base, to which is bolted a large dynamo-electric machine of the Siemens type, placed, however, vertically, or on end, instead of in the usual position. This allows free examination of the commutator being made. The steam engine is of a new rotating type, known on the Continent as Dolgorouke's patent. It has a double cylinder, with two rotating pistons. The velocity of rotation in this steam engine is so great that it may be connected direct to the axis of the machine. A small fly-wheel is, however, attached to the shaft to balance the motion. The horizontal steam engine driving the current generating machines for the Lichterfelde Railway belongs to the pumping station of the Lichterfelde water supply, and is only used temporarily.

FIG. 28.

It is found that when two dynamo-electric machines are connected up in parallel circuit, or as one machine, as directed in a previous chapter, a gain of from 20 to 30 per cent. over the total force given by the two machines when working separately is secured. By these means the currents for the electric cars can be obtained at a lower rate. Two machines are used to supply the current in the case of the Lichterfelde line.

On the continent of Europe alone there are at the present time many other examples of electric railways, and arrangements are being rapidly pushed forward for more than one permanent line of considerable importance, but the limits of the present work preclude the giving of descriptions of other examples. The Lichterfelde line being the first electric railway brought into practical use for ordinary purposes may also be assumed to present the best known means of constructing and working such lines.

While on the subject of electric railways, M. Achard's electro-magnetic wheel-brake may be mentioned. It is more especially applicable to ordinary railways. The apparatus attracted much attention at the Electrical Exhibition. It consists of two iron rings, about two inches wide, attached to the axle of the carriage. A separate axle also carries two iron rings of larger diameter and facing the first. The separate axle and rings may be transformed into a powerful electro-magnet by a current generated on the locomotive, so that the four rings may be attracted together with great force when the current is turned on, thus retarding the rotatory motion.

CHAPTER IX.

Minor Applications of Electro-Motive Energy.

The electro-motor has been applied to a considerable number of purposes of a minor description, with marked success. In the greater number of these instances other forms of motor would prove troublesome or inadmissible. There still exists a wide field for the development of the application to general purposes of small electro-motors, chiefly actuated by currents derived from the consumption of zinc, or from a thermo-electric battery, or from stored electricity. Miniature apparatus of this kind may be made to yield surprising results, within the most restricted limits imposed by the necessity for portability and lightness. In many instances the driving force may be and is obtained from a two-cell voltaic generator, or from a small accumulator, charged with some constant cell such as Daniell's. In some instances the current is produced by a thermo-electric battery, actuated by gas, on the principle adopted by M. Clamond and others. It is impossible to detail the various purposes to which these small motors have been applied, and the still larger number to which they might, with advantage, be adapted, but a few instances may serve to demonstrate the classes of work to which they are best suited, and in the practice of which they may prove most economical. It will be generally conceded, that even by the application of electricity derived from the consumption of zinc, the small moving powers needed in many kinds of work would prove inexpensive when produced by electro-motors. In cases of this description, the fact that a zinc motor is, power for power, twenty times as expensive in use as a coal-motor, will weigh very little when it is considered that coal or even gas motors are totally inapplicable, and that the great convenience of current power cannot be secured in any other way than by the employment of electro-motors. It may, indeed, be objected regarding a voltaic battery that it is troublesome, but this cannot any longer be adduced as condemning the use of small motors, since, by the invention of accumulators, a constant form of cell may be used to provide a store of energy, which may be used as required. Moreover, the thermo-electric batteries of MM. Clamond and Noë are available in many instances. By either of these methods a current may be obtained without much trouble, and at insignificant cost. But it will be seen, from the nature of the instances of the application of small motors adduced, that both trouble and cost weigh little, as already said, in the balance against convenience, applicability, and safety.

Electric Routing Machine.—The process of "routing" is that of clearing out and deepening the hollows or "whites" in wood engravings and in electrotype plates. After an electrotype is mounted, it is usually found that the whites of the plate are not sufficiently deep to allow of clearance in the printing. It is therefore common to chip or otherwise deepen them by hammer and chisel, or by what is known as a routing machine, driven by steam power. A rapidly-rotating cutter is brought to bear upon the work in this case.

The electric apparatus is intended to supersede both the old method and the steam-driven machine. It consists of a miniature electro-motor, carrying a rotating cutter, and driven by a current produced by any of the above generators or means.

Its constructional arrangement is as follows: Between the limbs of a permanent steel magnet, of the form represented in Fig. 14, and 6in. in length, is mounted, in the direction of the length, a small armature of the Siemens type, as represented in Fig. 11. The armature is free to rotate between the limbs of the field magnet, and in close proximity to them. It is furnished in the usual way with a reversing commutator. Its length, including the axis of rotation, is 4in., and its diameter, through the iron body, from pole to pole, 1in. It is excited by a coil composed of four layers of No. 24 (Birmingham wire gauge) silk insulated wire. This coil is carefully insulated from the body of the armature by the usual means. At either end the bearing for the axis is of gun metal, attached to the field magnet. The commutator is carried by an extension of the axis, as usual, and the feeding brushes are fixed in brackets insulated from the bearing by a block of ebonite. The opposite extremity of the axis is prolonged outside its bearing through the bend of the field magnet, and upon this end of the spindle is fitted a sharp miniature cutter. The whole apparatus is arranged in a similar manner to that represented in Fig. 15, but only one magnet is used. Its main portion is enclosed in a polished wooden case. As the whole weighs about 2lb., it is supported by being attached by a hook to an indiarubber cord capable of extension, and attached to the roof over head. By these means the weight of the apparatus is entirely taken off the hands, which are thus left free to guide the revolving cutter in any desired direction with great nicety. The electrical resistance of the motor is 0·6 ohm.

The necessary conductors from the electric source are brought down within the suspending rubber cord, and attached to terminals fixed in the upper end of the motor. In order that the double conductors may not interfere with the expansive action of the supporting rubber cord, they are connected to two expanding brass wire spirals at the upper end. A circuit interrupter, in the form of a finger lever at the upper end of the motor, enables the operator to start and stop the motor at will, by opening or closing the circuit. The velocity of rotation of the cutter is very great, and may probably be taken as equal to 3,000 revolutions per minute when the armature is of the softest iron.

It would appear to be in some respects more advantageous to employ an electro-magnet instead of a permanent magnet to set up the magnetic field, for although the apparatus would in this case be necessarily more complex in construction, and would probably require a little addition to the actuating current, its weight would probably be reduced to about one-half. In such case a miniature model of the electro-magnetic field motor, described in Fig. 16, with its magnet excited by a coil of No. 24 wire, would answer the purpose. In this case the polar inductors between which the magnetic field would be set up should be either of hard cast iron or steel. The reason is to be found in the tendency of small masses of soft iron to rapidly become inert when the exciting current is interrupted; and it is probable that even the tendency to interruption created by the reversal of the current in the armature would greatly weaken the arrangement, unless the polar pieces were of a nature hard enough to retain a certain store of magnetism permanently.

Electric Stone-engraving Apparatus.—An arrangement somewhat similar to the above has been tried on a class of work known as stone dressing or engraving. The motor is about one-third larger, its dimensions being as follow: The inducing magnet is 9in. in length, from between the poles to the middle of the bend. It is of the compound kind, made up from four thin magnets of the same size, fastened together with screws. The armature is of the Siemens type, 5in. in length, and with a diameter from pole to pole exteriorily of $1\frac{1}{4}$in. The inducing wire is of No. 20 gauge, silk-insulated, in four layers. The resistance is about 0·5 ohm only.

The rotating spindle is fitted with a reversing commutator made of copper, the contact springs being of steel, faced with platinum, a metal so difficult to fuse that it suffers but slight deterioration from the sparks given off at reversal and partial interruption of current. The opposite extremity of the axis of rotation carries a cutter adapted for engraving stone. The velocity of rotation varies with the work from 1,000 to 3,000 revolutions per minute. This motor may be actuated at sufficient speed by a current equal to that given by two small cells of the Bunsen type.

The apparatus may be guided in the hands, or supported by a suspending indiarubber cord, as in the former case. The cutter can be applied to the work with great nicety, and its rapid rate of motion causes the softer varieties of stone to yield with surprising effect. Geometrical patterns may be described by this means, lettering and other designs may be cut, and other operations effected with greater speed and less risk of damage than by means of chisels and hammers. A machine with a chisel or cutter having a reciprocating motion is found to act upon stone with great effect. The above motor may be used for this purpose by fitting it with bevel-wheel gear, so arranged that the rotatory motion may be transformed into a reciprocating movement of a steel bar moving in a vertical line. An eccentric or cam on the second mitre wheel may be used for this purpose. By means of a motor with a reciprocating armature devised by the author in 1879, it was found that a speed of 2,000 effective strokes per minute might be imparted to the cutter. A motion through three-eighths of an inch is sufficient.

It would probably be better in a case of this kind to dispense with the usual rotating armature altogether, and to employ instead one of the attracting and repelling coils or solenoids described in Chap. V. In an arrangement of this kind a bar of iron or magnetised steel might be made to acquire a very rapid rate of motion through a short distance. This device would also offer other advanvantages. The stroke given when an eccentric or crank is employed is quite positive up to the point of reversing the motion. It is thus limited, which might unfit the apparatus for working upon certain kinds of material. The stroke given by a plunger in a solenoid is not limited or positive. In this respect it resembles a spring, which may be more or less compressed. Thus an apparatus for stone-dressing and other operations, even rock-drilling, might be devised, and such an arrangement would present many features of practical advantage.

Mr. Cranston, a mining engineer of Newcastle, was the first to propose to employ electricity for actuating rock drills. The ordinary rock drill consists of a vibrating piston in a cylinder, the motive power being either steam or compressed air. To the lower end of the piston rod is attached a flat-faced or other chisel, which, being caused to repeatedly strike a particular part of the stone, is slowly rotated by hand, the result being that the hardest rocks rapidly yield to the continuous breaking-up action of the drill or cutter. In like manner the electric impulse employed by Mr. Cranston, and also lately for the same purpose by Messrs. Siemens, of Berlin, causes the drill to vibrate at a prodigious rate and with great power. The current of electricity is derived usually from a dynamo-electric machine, but the accumulator might probably be brought into requisition for this purpose with less cost. The advantage of this novel application of electricity lies in the applicability of the current, and the ease with which an electric conductor of any length may be carried to the work, in positions frequently inaccessible to steam or air pipes. Moreover, compressed air and steam applied in this way are enormously expensive—the former being costly on account of the great loss of energy involved in its compression, and the latter on account of the waste of heat and power involved in its conveyance through great lengths of tubing. Electricity, on the other hand, may be generated at any convenient distance without considerable loss, or it might be stored in accumulators, which might be brought quite near to the drills, the loss in either case, due simply to conductor resistance, being comparatively insignificant. The current from one source of electricity, as a dynamo-machine, may in this manner be employed to actuate a considerable number of rock drills in one circuit simultaneously.

Millstone dressing, a process involving either costly diamond drilling machinery or skilled labour, may also be effected by means of apparatus actuated by the motive power of electricity. It would appear that in this case both vibratory and rotatory motions might be used with advantage.

Dentists' Drills.—The electric current, from the facility with which it can be conveyed through a thin flexible conductor, is eminently adapted for actuating the apparatus employed by dentists. By the usual method a dentist's "engine" is set in rotation by the foot, and its motion is communicated to the drill by means of a long flexible steel spiral, but this arrangement has many disadvantages, since it either necessitates the employment of manual power, which is generally effected by means of a treadle worked by the operator, or a gas motor. A drill worked by clockwork was invented for this purpose, but it proved too bulky and noisy. An electro-motive drill is entirely free from these disadvantages, and it may easily be made small enough to obviate any objection on that score.

The electro-motor may be made of the following dimensions, and yield an effective drilling power. The field magnet should be of the electro kind, as adapted for the type of motor exhibited in Fig. 19. This form is very compact. It may be 3in. in length. Its exciting coil may be composed of four layers of No. 26 silk-insulated wire. The miniature Siemens armature may be $2\frac{1}{4}$in. in length, including the axis of rotation. It should be so mounted in bearings of gun metal as to freely rotate between the polar inductors, which in this case may be of steel, with the bearings secured across their ends. If thought desirable, the motion, instead of being taken direct from the axis, and in a line with it, may be turned off at right angles by means of suitable gearing. The fittings and necessary commutator should be of the kind already described, the contact springs of the latter being of steel, faced with platinum. A permanent magnet, made up of several thin magnetised steel *laminæ*, may be employed, but the apparatus would probably, in this case, be more bulky and of greater weight. The conductors should be of the kind used for telephones—two thin flexible wires twisted together. A very small current would suffice to actuate this motor at a very great speed, with a yield of considerable drilling power. The weight of an arrangement of this kind might probably be reduced to 4oz.

Electric Pen.—This little instrument is due to the ingenuity of Mr. Edison. It consists of a small two-limbed electro-magnet, in front of the poles of which rotates a flat iron armature. Its construction is essentially the same as that exhibited in Fig. 7, except that an electro-magnet takes the place of the permanent magnet, and the armature is a plain piece of iron. The current is cut off at the instant when the armature coincides with the poles, and the armature continues its rotation by inertia. A miniature three-throw cam is fitted upon the axis, by means of which a thin steel rod, fitted at its lower end with a piercing needle, and working through a steel sheath, is caused to vibrate three times at each revolution of the motor. The rate at which the needle vibrates in the sheath is enormous. Its piercing effect is utilised by forming written characters, in a line of closely-connected perforations in paper. This paper is afterwards used as a stencil plate, through which any number of impressions, in ink, may be obtained. But the method has practically been superseded by the "multigraph" and other processes, more generally applicable, and not involving the use of electricity.

Rotating Engine for Vacuum Tubes.—Small motors, for the purpose of giving a rotatory motion to the vacuum tubes used for exhibiting the effects of electric discharges in rarified gases and air, have been made according to various designs. Probably the most compact, effective, and inexpensive motor for this purpose is that represented in outline in Fig. 29, which exhibits the essential portion of the apparatus. It will be observed that a straight electro-magnetic bar, with its exciting coil in two sections, is used as the rotating portion of the arrangement. This electro-magnet is wound with a direct helix of insulated wire, so that, upon passing a current, its extremities assume N and S polarity respectively. The straight electro-magnet is mounted, centrally of its

FIG. 29.

length, upon an axis, rotating in bearings attached across the armature or surrounding circular portion of the motor. This armature is of peculiar shape. It is a ring-shaped armature of soft cast iron, with segments cut from its interior, as represented. This arrangement presents a succession of alternating elevations and depressions, so as to afford a series of attracting points and neutral points to the force of the magnet as it rotates.

It will also be observed that the shape of the attracting portions is favourable to the production of a kind of progressive and prolonged pull upon the electro-magnet, by which means it approaches its poles closer and closer to the armature, until the extreme point is reached, when the current is entirely cut off. The circuit is thus completed and interrupted six times during each revolution.

The circuit breaker or commutator is attached to and revolves with the axis. It takes the form of a copper wheel, with six teeth or elevations on its periphery to represent the six interruptions necessary. It is insulated from the axis. The contact spring is also insulated from the metalwork of the machine. One extremity of the magnet coil wire is attached to the framework, forming one of the poles. The other extremity is connected to the insulated contact wheel. The contact wheel is so set as to allow the spring to touch one of its teeth, and so allow the current to flow as the electro-magnet approaches an elevation on the armature. The cut-off takes place just as the magnet is passing the elevation. The rate of rotation may be made rapid or slow, by applying a large or small current. One small cell maintains the engine in motion with sufficient power and speed.

The vacuum tubes are usually attached to an arrangement carried by the axis of rotation. This consists of an arm or beam attached at its centre to the axis and insulated from it. The beam consists of pieces of ebonite and brass, so arranged that the secondary current used in illuminating the tubes can be passed through them as they remain fixed between two spring catches carried by the beam.

The frame of the motor is thus utilised as the return portion of the circuit for both rotating and illuminating currents, and two pairs of terminals are provided on the base of the machine for the two currents.

Electro-Motors for Sewing Machines.—A considerable number of years ago the proposal to run sewing machines by means of electro-motors attracted attention. The Howe Machine Company brought out a small motor similar in principle to Froment's, as exhibited in Fig. 10. This small engine served the purpose so far as the necessary speed and force were concerned, but its use necessitated the employment of a considerable force of current, which was not easily obtained save by means of voltaic cells, and the motor therefore never came into use. The effective motive power was about 1,000 foot-pounds, and to produce this a current equal to that from seven Bunsen cells in series had to be passed through the motor. The same result may be obtained from the improved motors with a current of one-half the strength.

It is by no means easy to say what force is necessary to run a sewing machine, since different machines, or different types of construction, necessitate the use of very different moving powers. Thus the motive power that would run a large Howe machine would probably suffice to actuate two or three of the lighter Wheeler and Wilson machines, while a very considerable power would be needed to work the larger manufacturing machines made by various companies. The average power necessary may probably be stated at 500 foot-pounds per minute. Under favourable circumstances this motive force may be obtained from a very small electro-motor.

The magnetic field for a motor of this description may be obtained by the use of a permanent magnetic battery of about twelve magnets, each 8in. in length from between the poles to the middle of the bend. These may be clamped in connection with a pair of polar extensions of cast iron, in the manner described in connection with Fig. 12. A still greater number of magnets will increase the intensity of the magnetic field and the power of the motor. It is essential that the magnets be of the best steel, and that each should be magnetised to saturation. In connection with these a Siemens armature, 12in. long in the web, may be used. This will give a total length of about 15in. to the complete machine. The axis of rotation should be of steel at either end of the armature, bearing in blocks of gun metal fastened across the cast-iron inductors. The coil on the armature may be of No. 16 cotton-insulated wire, in four or five layers. The motor would prove more effective if provided with an electro-magnetic field, as in Fig. 17. In this case the electro-magnet should be composed of $\frac{3}{8}$in. boiler-plate, curved to the U form required to bring its poles to embrace the exterior sides of the magnetic inductors. The core, after being bent, should be carefully annealed. In the processes of finishing care should be taken to remove all roughnesses in the course of the wire, and to chamfer or round off the sharp angles or other points likely to chafe the insulating covering of the exciting coil. The wire should be of No. 14 gauge, and it may be covered with cotton. Insulated wires in cotton, silk, and guttapercha are readily procurable of dealers in electrical apparatus. Four layers should be wound on the core. In other respects the descriptions previously given will suffice to show in what manner the other portions should be made and fitted together. A motor of this description and these dimensions yielded an effective power sufficient to actuate an ordinary sewing machine, with a very moderate current.

The motor may be attached to any convenient part of the sewing-machine stand. To actuate a Wheeler and Wilson machine of the No. 1 or 2 description, it should be arranged underneath the table, with a band to the pulley of the machine. A contact or circuit maker should be attached to the machine, so that the motor may be started or stopped by the motion of the treadle without the intervention of the hands.

It is doubtful whether there exists any considerable demand for a motor to actuate sewing machines as ordinarily employed. The greater number of workers would probably prefer to actuate the machine by foot, more especially since power of any kind must add greatly to the cost of producing machine sewing. In many special cases, however, when the machine is not used for direct profit, an electro-motor may be employed with the most pleasurable results.

Transmitter for Postal Messages.—Dr. Werner Siemens has suggested that the electro-motor might be utilised with advantage on small railways for the transmission of postal messages, analogous to the pneumatic-tube system already in use, but available for longer distances than the latter, and for "elevated railways." For postal service he proposes a small covered railway track. about 18in. high, mounted on short iron columns, erected alongside of existing railways, or by any other route. The two rails are insulated from each other by being mounted on wooden sleepers—one being in connection with the sheet-iron covering, the other in connection with the earth by means of the iron supporting columns, to form the conductors for the current generated by the stationary machine. The circuit is established with the motor on the axle through the wheels by a suitable commutator. The electrical resistance of such a system of conductors would not amount to more than 0·030 ohm per mile, so that one stationary machine would suffice for a length of about fifteen miles. If the stationary machine were of such a construction as to give a considerably more powerful current than the motors, the speed of a carriage would not be sensibly diminished if others were introduced on the rails at the same time, it being a property of the dynamo-electric machine that the power increases at a greater ratio than the speed of rotation. A speed of over 1,000 yards per minute could, Dr. Siemens thinks, easily be obtained upon such a miniature railway.

Messrs. Siemens Bros. employ the electro-motor to work lifts or hoists. A small machine is fitted in the hoist cage, which is balanced by a counter-weight, and causes it to ascend by gearing into a rack or screw on the fixed rail, from which, also, the current from a dynamo-electric machine may be obtained. This appears to be a most suitable application of the electro-motor, and it will probably be extensively employed in cities, in cases where dynamo-electricity is available.

M. Trouvé employs small electro-motors in the propulsion of light pleasure boats, the source of electricity being carried by the boat itself. M. Trouvé has also made experiments in the use of motors for propelling tricycles. He applied the power of a small motor to each wheel of the tricycle, and was enabled to propel the machine through the streets of Paris, at a speed of ten miles an hour. In this case the current was obtained from a Faure accumulator, carried by the machine itself. The Faure cells were previously filled or charged by being connected to a dynamo-electric machine.

The same type of motor is especially suitable for use in the laboratory and lecture-room, for turning electrical machines, experiments in the recomposition of light, etc. The motors may likewise be utilised at small expense for the working of fans, punkas, ventilators, blowing organs, working clock-maker's tools, and many other kinds of light work.

In many cases the necessary currents may be obtained from already-established electric-light circuits. In others, from small voltaic cells, free from fumes, and in most cases from accumulators, in which electricity has been stored by machine, thermo-battery, or cells.

Minor Sources of Electricity.—The three leading sources of electrical energy at present available are *heat, chemical decomposition,* and *mechanical motion.*

The first is at present represented by thermo-electric batteries, the second by voltaic cells, and the third by magneto and dynamo electric machines. Each of these may be said to be applicable to the purpose of furnishing energy for driving electro-motors, according to the extent of the power required. Thus, present thermo-electric batteries may be employed in the turning of small electro-motors for working ventilators, punkas, or sewing machines. The voltaic cells, in which zinc is consumed, and indeed represents the fuel, are applicable to the same purposes, or when a stronger current is required. The third is more especially applicable in the transmission of power from one point to another, or in the distribution of power to different points from one source of electricity.

In saying that thermo-electric and voltaic cells may be used for driving electro-motors direct, we have represented only one aspect of this question of small powers, and that the most unfavourable one. Of late, means and apparatus for *storing* or *accumulating* electrical energy have developed into inventions of great importance and commercial value. Prior to the discoveries of MM. Planté and Faure there existed no practicable means by which electricity could be accumulated or stored. Therefore, the smaller or feebler generators of electrical energy were in most cases valueless in regard to the driving of electro-motors. The secondary battery or accumulator possesses a double value in relation to small sources of electricity. It will not only condense the energy in quantity, but will give it up, either in quantity as stored, or in the form of a current of great *electro-motive force*. To render the case clearer, suppose it were required to utilise to the full the energy of a five-cell voltaic battery of the Daniell type. Let this battery be connected direct to an electro-motor and the moving power would be very small. Let us call its electro-motive force five volts. Suppose the battery to be connected to a series of ten accumulator cells, arranged as one cell of large surface. Under these conditions the Faure battery would be charged in quantity to a potential of two volts. Still, even when fully charged, the electro-motive force would only be two volts from the Faure battery. But let us suddenly join up the Faure cells in *series*, instead of multiple arc, and the current obtained will be intensified or reinforced *ten times*. The accumulator will now furnish a current of *ten times the electro-motive force of the Daniell battery*. This powerful current will continue to flow until the accumulator has come to rest. In the nature of things the intensified current cannot flow so long as that of the Daniell cell while it charged the accumulator. If, for example, the Daniell battery were allowed to charge the accumulator from four o'clock in the day to twelve o'clock next day, the charging, supposing the accumulator to be sufficiently large, would last twenty hours. This would enable us to obtain, by throwing the accumulator cells into series, a most powerful current, ten times the strength of the Daniell battery, for two hours. Or if we arranged the accumulator to give half the electro-motive force, the current would be five times stronger than the Daniell current, and would last four hours. By these means it is doubtless practicable to make small currents of electricity extremely useful. There is nothing paradoxical in the method. It is simply one way of utilising the nature of the conservation of energy. The current is ten times strengthened, but it flows for one-tenth of the time. It is analogous to the effect of employing a small motor to wind up, to a given height, a series of 1lb. weights, each of which is released or deposited at the given height and then left until wanted. Thus at the end of many hours many tons might be accumulated at the given height. This accumulation of power might then be employed to do small work throughout considerable time, or to do a great deal of work in a short time. From this point of view it is impossible to overestimate the importance of an electrical accumulator. More especially is this true in relation to the many sources of comparatively weak currents at present available.

Considerable attention and ingenuity have of late years been devoted to the subject of thermo-electric generators, or converters of heat into electric energy. The result is that a very considerable proportion of the heat energy employed may now be recovered as electricity, and comparatively small apparatus suffices to yield a current of considerable power.

Several inventors have given models of thermo-electric batteries, the most successful of which evince inventive powers and scientific knowledge of the highest order. Those of M. Clamond and M. Noë are especially worthy of mention. In cases where heat is obtainable at a cheap rate, as obtained from charcoal and gas, the thermo-electric piles, as they are usually termed, may be used with advantage, to furnish constant currents of moderate power. But thermo-electric batteries have not yet been constructed of sufficiently large dimensions to yield the powerful currents, for direct use, required to drive machinery. The sizes that may be obtained commercially are generally heated by a few gas jets, or in larger models by charcoal. Some of Clamond's 100-bar piles are stated to yield a current equal to 4 webers, the mechanical equivalent of which would be, according to Dr. Joule, about 222 foot-pounds per minute. The expenditure of coal gas to obtain this current is stated to be about 10 cubic feet per hour. The larger thermo-electric batteries are heated by means of a charcoal fire, as above mentioned, but they are not so extensively used as the gas-heated piles. One objection which is stated to apply to thermo-electric piles, namely, a decrease in the current yielded after long use, would appear to be due to an increase of the internal resistance. This fault is said to be entirely eliminated in the construction of the piles issued by M. Noë, who employs an alloy different from that used by M. Clamond. Very considerable improvements have, however, been made of late in all converters of this class, and it may be assumed that a thermo-electric battery of such construction as will insure a current of constant strength is not now difficult to obtain. The chief objection would appear to be the first cost, which is as yet considerable. There is every prospect that the expensive materials at present employed in their construction will be replaced by equally effective combinations of metals and alloys of a less costly kind. Any kind of fuel, such as the ordinary illuminating oils, may be employed in lieu of gas or charcoal; waste heat as steam is also available. An illustrated description of the more effective thermo-electric piles will be found in the author's treatise on "Electrotyping."

Voltaic or Zinc Generators.—The number of

different kinds of voltaic batteries is very great. Only a few are, however, suited to the work of actuating electro-motors direct. In fact, were it not for the facility with which electricity can be accumulated, condensed, or stored, voltaic converters would in most cases be useless for actuating electro-motors. Some of the weaker cells may now be used to charge accumulators, and in many cases the cost of zinc is, compared with the convenience attending the use of an electro-motor, trifling. Voltaic batteries may be constructed of any required dimensions, but it is improbable that they will ever be used to produce currents of considerable power direct.

From these considerations, and the fact that many special applications of power have of late arisen, owing to the great convenience of electro-motors, the voltaic current will be considered as offering many points of advantage, even although it should cost a high price as compared with motive energy derived from coal. Our chief object in the present section of the chapter is to obtain a clear view of the production of voltaic currents under conditions when all waste is reduced to a minimum.

The zinc generator may be regarded as a kind of furnace where the fuel is zinc. It is important to take this view of it at the outset, because the current is derived from the potential energy of zinc in the process of its dissolution and combination with oxygen to form oxide of zinc, just as the steam engine derives its motive energy from that of coal combining with the oxygen of the air. Considered thus, the voltaic furnace can be shown to be a much more perfect and specifically economical arrangement than the steam furnace. All heat of low grade is wasted or lost in the steam furnace : in the electric generator all potentiality is utilised.

It is an interesting fact that when a piece of zinc is plunged in acidulated water it begins to dissolve and give off *heat*. It is still more important to know that by providing certain conditions, in the shape of a negative or passive plate of copper or carbon, connected to the zinc with a wire, this heat will diminish, and it will be transformed into a current of electricity in the junction wire. This current may be conducted to any point, there to yield its energy as mechanical motion. Almost all the specific energy of the dissolved zinc may thus be obtained in properly-designed apparatus. A voltaic converter should be so arranged as to fulfil the following conditions :—

1. The construction must be such that it may be conveniently shifted from place to place. It must be compact, and of small size in proportion to its power (portable).

2. Generally it must not emit any offensive fumes or odours, and it must be adapted to consume zinc under the most favourable circumstances.

3. Its design must allow of the free examination of its constituent parts, and facilities must be provided for the renewal of the zinc plates and excitants with the minimum of trouble.

4. Electrically considered it must have a low internal resistance, in conjunction with a high electro-motive force.

5. It must not frequently require attention; for several hours at a time the current must be of con-

M

stant strength, and during that period there should be no necessity for attention.

All these advantages may be combined, but in the common forms of the voltaic cell it is useless to seek for them in combination. It may be useful, however, to become acquainted with one form of the voltaic cell as hitherto employed.

One of the most commonly useful zinc converters is that known as Bunsen's cell. It has been very extensively employed for various purposes requiring the expenditure of large currents. It possesses a high electro-motive force (1·888 volt), a small internal resistance, and its energy is almost constant for several hours at a time. It may be arranged to act with different kinds of excitants, and so to yield various electro-motive forces, through various resistances. When arranged to act slowly, with a weak excitant, the current flows for days at a time without the necessity for attention. The first cost of the cell is small, the materials of which it is composed being in common use and everywhere obtainable. This cell presents other points of advantage over many other voltaic generators, but those mentioned will suffice to show why Bunsen's cell has been so extensively used.

Two vessels and liquids are used as in Daniell's cell. By these means the hydrogen evolved by the absorption of oxygen to form oxide of zinc is either absorbed again before it reaches the negative plate or is set free from the negative cell as a constituent of hyponitrous acid gas. When the cell is working at its strongest some of this gas is given off as fumes, which are both disagreeable and unwholesome. When a smaller proportion of hydrogen is evolved at the zinc surface it is wholly absorbed in the negative cell.

In the common form of the element an outer containing vessel of glass or earthenware is employed. The inner vessel is much narrower, and of unglazed earthenware, forming a *porous* cell through which the current can pass, but which suffices to separate the two liquids of the element. The cells may be of any required shape. It is common to use cylindrical vessels, the dimensions of the outer one determining the size of the cell. Its capacity varies in the common sizes from a pint to several gallons. The positive and negative plates are of zinc and gas carbon. The zinc is usually in the form of a cylinder, amalgamated, and composed of rolled Belgian plate. A terminal is attached to the zinc. The negative plate is usually in the form of a rod or rectangular block of the graphite variety of carbon found in the interior of retorts after the process of gas-making. It is not coke, although the two varieties bear a resemblance to each other. A suitable quality of the carbon presents a clear grey appearance, and is extremely hard. Difficulty is frequently experienced in obtaining an effective electrical connection with the carbon rod. A brass or copper clamp, screwed tightly upon it, affords a sufficient connection for short periods of working, especially when the clamp can be removed and frequently cleaned. A more effective connection is obtained by tipping the contact screw with platinum, which is not corroded by the fumes. But all the ordinary methods of joining the wire to the carbon fail sooner or later. The only permanently effective method consists in depositing a copper cap upon the extremity of the carbon. This may be done by cutting notches on the head, drilling a hole, and fastening a stout copper wire through it, to afford a better hold for the head, and depositing a copper cap, to the thickness of brown paper, in a Daniell cell. It is only necessary to tie the wire leading from the zinc of the cell around the carbon, and to dip the head to be coppered in the sulphate solution of the outer cell, having previously removed the copper cylinder. A deposit of copper at once begins to form, and in a day or two will have attained the desired thickness. When ready for removal, a hole or two should be drilled through both copper and carbon, in order to allow of soaking out the copper solution in hot water. Afterwards dry by heat, and apply solid paraffin, having previously soldered the terminal screw to the cap. The solid paraffin is intended to prevent the acid from ascending and destroying the junction. The liquids which may be employed as excitants for the Bunsen cell are very numerous. The electro-motive force and resistance vary with the kind of liquid employed. When the zinc compartment is nearly filled with a mixture of one part sulphuric acid to twelve of water, and the carbon compartment (porous cell) with strong nitric acid, the electro-motive force is at a maximum (1·964 volt), and the internal resistance at a minimum (ordinarily from $\frac{1}{8}$ to $\frac{1}{4}$ ohm, according to the size of cell). When the cell is arranged in this way it yields nearly the greatest effect obtainable from a voltaic generator. The internal resistance being assumed at a quarter of an ohm, and the electro-motive force 2 volts, the current in an interpolar wire, presenting a resistance of $\frac{3}{4}$ of an ohm, would be two webers. The mechanical equivalent of this force would be, according to the experiments of Dr. Joule, nearly 89 foot-pounds per minute. The corresponding consumption of zinc, according to the law of equivalents, would be about 35 grains per hour. After a few hours, the gradual transfusion of the separated liquids, better known as endosmose, begins to tell upon and increase the resistance, and the strength of the current slowly falls off. The liquid used in the porous cell may be used several times. By degrees the absorption of the hypo-nitrous acid gas causes it to assume a red colour, then a green tint, and finally it appears as a colourless and inert liquid, differing little from water.

The same type of cell may be charged differently, and different results follow. When the nitric acid is represented by a mixture of twenty-five parts of the acid in a hundred parts of water, and the excitant by nine parts sulphuric acid in a hundred parts of water, the electro-motive force will be 1·63 volt. The internal resistance will, for the same size of cell, be about one-eighth of an ohm greater. When the cell is thus arranged, the current, under the before-mentioned conditions, will be about 1·7 weber, but the periods of activity of the cell will be considerably prolonged. Thus arranged the cell will not emit offensive fumes, and there will be less waste of zinc from local action. A still weaker excitant may be used. The nitric acid mixture may be the same, and a solution of common salt may be used in the zinc compartment. The electro-motive force in this case is 1·9 volt, but the internal resistance is increased, so that no advantage, except that of still further prolonged activity, is gained. By charging the carbon compartment with a solution of 2oz. of bichromate of potash in 20oz. of water, acidulated with 10oz. of sulphuric acid, and employing, as before, common salt in the zinc compartment, an electro-motive force of 2 volts is obtained, with a moderate resistance and no fumes. The constancy of this cell is very remarkable, but its resistance is frequently too high for direct use in electro-motor circuits. The arrangement is very suitable when it is desired to store the current. The resistance is considerably diminished by the use of acidulated water in the zinc compartment.

A simple cell, from which a current of great strength may be obtained for a short time, may be constructed by employing two plates of carbon and one of amalgamated zinc placed between them. The three plates are usually attached to a beam of wood, and so insulated. The two carbon plates are connected together as one by a brass clamp. This combination is excited by a solution of bichromate of potash, acidulated by 2oz. (to each pint) of sulphuric acid. The resulting action is exceedingly vigorous, but it is not maintained unless the plates be disturbed so as to cause circulation of the liquid.

Various attempts have been made to cause the solution to circulate of itself, and so prolong the action of the cell. The best of these is the sustaining battery devised by the author. It consists of an arrangement by means of which gravity causes the exciting liquid to continue in motion until it is exhausted. Fig. 30 represents an element made according to this device. The containing vessel is cylindrical and deep. The plates are two discs of carbon and one of zinc, placed in a horizontal plane, in a wooden case fitting the containing vessel in the manner of an air-tight piston. The piston is packed by means of a rubber ring, stretched in a groove turned in the face, as represented in section. The carbon plates may be connected together by

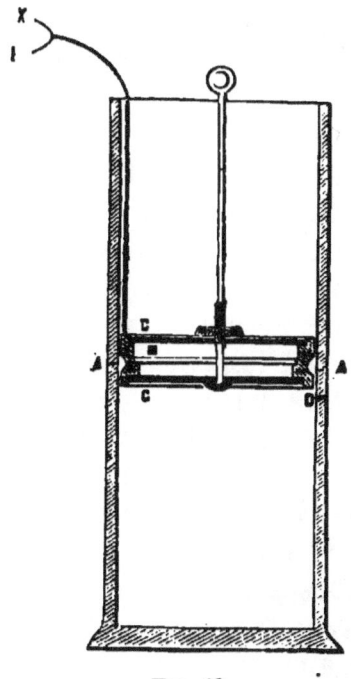

FIG. 30.

means of a metallic stem, which is free of the zinc by an aperture cut in the centre. This metallic stem must be of platinum. Ivory or ebonite may be used, and the connection may thus be made between the carbon plates, and to the exterior conductor, by means of a piece of platinum wire in the wooden annular frame. The connection with the zinc is made by means of a plug and taper tube of platinum, with a guttapercha-covered wire leading out of the cell. By these means the zinc plate may be replaced by a fresh disc when worn without trouble—no soldering is necessary at any part of the element. The lifting stem may either be of two parts, so as to divide, or it may be made to act as a nut to screw down upon the carbon plate. The exciting liquid is poured into the upper half of the containing vessel, and flows into the element through an aperture in the upper carbon. A still smaller aperture is provided through the lower carbon plate, and the liquid thus slowly percolates through the cell, falling into the lower compartment drop by drop. A small air outlet is provided at the upper end of the lower compartment. By these means the action of the cell may be continued, in full even flow of current, as long as any of the exciting liquid remains in the upper compartment. It is thus possible and easy to obtain from the bichromate cell more favourable results than from the Bunsen generator. The force is greater, there is no fume, the resistance is much less, and the constancy is superior. A cell of the above type may be made to continue in action for days together, and supply its full force throughout that time. When the lower compartment is full, and the solution be deemed still strong enough, the element may be slowly depressed. This will force the lower liquid up through the element into the upper compartment once more.

One of the great advantages of this type of element is its extremely low internal resistance, which is often not more than 0·1 ohm. The plates may be placed very close together, compatible with allowing due circulation of the excitant. The zinc plate should in all cases be carefully amalgamated. The current yielded is frequently as great as that from two Bunsen cells of equal size.

Grouping the Cells.—When the carbon of one cell is connected to the zinc of the next, and so on through the series, it is said that the cells are joined up "in series" or for electro-motive force. By these means we increase the electro-motive force in the ratio of the number. Thus, two cells in series will exhibit an electro-motive force twice as high as that given by one cell. When all the carbons are connected to one conductor and the zincs to another, the cells are said to be connected in parallel circuit, or simply as one cell of large surface. In this case the electro-motive force is that of one cell only, but the quantity of electricity given off is proportional to the number. In the case of joining up for force, we increase the internal resistance of the battery in the ratio of the number. Two cells joined for electro-motive force are capable of overcoming greater interpolar resistance than the same cells connected in parallel circuit. In the greater number of instances it will be found that about 4 volts is the minimum of electro-motive force by which motors may be actuated direct. In the charging of accumulators or Faure cells, the force may be small or great, according to the time expended in the charging. For motors of moderate power, a convenient force is from 10 to 15 volts, in a total resistance of about 2 ohms. From experiments it was found that the current in webers passed through a medium-sized Siemens machine was, for $2\frac{1}{2}$ actual horse power excited, 31·9, reclaiming a percentage of 47·7 of the prime power, through a resistance of nearly 0·5 ohm.

Size of the Elements.—To those who are likely to employ the voltaic elements for direct driving, or in the storing of electricity, or for testing purposes, it will prove useful to know that the electro-motive force of the element is not affected by the size. Thus a pair of thin wires used as a cell would yield the same electro-motive force as a pair of plates of the same metals. But as the element is increased in size, its internal resistance is diminished in direct ratio or nearly so, the result being the flow of an increased quantity of electricity. Thus, a large cell is undoubtedly a more powerful arrangement than a small cell, but under certain conditions it would not give a greater current. When the interpolar resistance is as small as the internal resistance of the cell, the useful effect is at a maximum; when the interpolar resistance is much increased, the useful effect from the same cell diminishes, and a smaller cell would then be just as effective. Hence the internal resistance of the cell and that between its poles are mutually correlated. Taking, say, twenty cells, each with an internal resistance of 0·1 ohm only, and each an electro-motive force equal to 2 volts, and connecting them up in a series so that the total internal resistance shall be 2 ohms, and the electro-motive force 40 volts, the current in an interpolar resistance of 2 ohms (total 4) would be be about 10 webers ($\frac{40}{4}=10$). But the greater number of circuits in small motors would begin to heat, and the electro-magnets would be magnetised to saturation by such a current, so that there is a limit to the current which may be economically passed through an electro-motor. The current must be limited, chiefly by reason of the fact that the electro-magnets would not utilise it beyond a certain strength, dependent upon their size. A limit is also imposed by the time required to magnetise and demagnetise the electro-magnets or armatures. It is nearly impossible to ascertain accurately what current would cause a given electro-motor to yield the maximum of power of which it is capable without actual trial, because the mass of iron to be magnetised and demagnetised, its specific tendency to retain magnetism, and its point of saturation, are all correlated, and vary to so great an extent in different motors.

The great general rule to observe is based upon the relative proportions of electro-motive force and resistance, since the former divided by the latter determines the current; and the current should be adapted to the motor, not only with reference to the resistance but with regard to the nature of the motor and its velocity of rotation. It may be generally assumed that when the interpolar resistance equals the internal resistance of the electric source, and when the motor does not tend to heat, the zinc is being consumed under economical conditions, or, in other words, the force is evenly distributed over the whole circuit.

Economical Working.—When voltaic cells are used to drive small motors direct, or to charge accumulators, certain conditions should be observed if it is required to gain the greatest effect from the minimum expenditure of material. In order to obtain from a given battery the full effect representing the combustion of zinc going on, care should be taken to reduce all the resistances to the lowest point, by placing the plates near to each other, and by making all connecting points and junction wires as large as convenient. The connecting-up arrangement should present so small a resistance that it may readily be left out of account. The zinc plates should be frequently examined, and re-amalgamated when they show bare patches. Cells should be thrown out of action when not in use, so that the zinc may not be wasted by local action.

CHAPTER X.

Fragmentary Information.

Working Cost of Sources of Small Currents.—
The small or miniature electro-motors yield a better
return for the energy expended than large motors.
This may be due to several causes. It may be a
result of less loss in the body of the armature, owing
to the greater rapidity of the reversals. This loss
in the larger machines may be assumed to appear
as heat. It is very probable that the small motors
return about 80 per cent. of the work of the
current.

The cost of electric energy, as derived from
sources consuming zinc, is always very high. It
varies with the cells from which it is obtained, the
Daniell cell being probably the most efficient
source, and the bichromate or nitric acid cells the
most expensive. If the energy of the current of a
Daniell battery, working as it does in the telegraph
form, can be stored up in a Faure or other accumulator at a low potential, and then released at a
high potential by throwing the secondary cells into
series, with a loss of 10 per cent. only, it would
appear that small working powers, applicable to
numerous purposes, might thus be produced at
small cost. The work of storing would be done
more rapidly by means of a sustaining cell (Fig. 30)
on the bichromate principle, but the cost would
be rather higher.

Dr. Joule calculated some years ago, before accumulators were invented, that under favourable circumstances, a current representing 1hp., or a current capable of actually developing that force through a magnetic motor, would necessitate the consumption of $37\frac{1}{2}$lb. of zinc in the Daniell battery in twelve hours. Taking this figure as representing the only substance consumed in the generator, at 4d. per pound, the cost is 12s. 6d. for one horse power maintained for 12 hours, or 1s. $0\frac{1}{2}$d. per hour.

From experiments made with Grove, Bunsen, and bichromate cells, in order to compare their working cost with that of the Daniell, it appears that the bichromate single cell costs nearly twice as much as the Daniell. The Bunsen cell was still higher. To maintain a weber current for 12 hours, the Daniell cost for zinc, 0·2d., the Bunsen, for zinc, nitric and sulphuric acids, 0·6d., and the sustaining bichromate 0·5d.

Dr. Joule gives the mechanical equivalent of the weber as ·735 foot-pound per second, or, say, 44·2 per minute. Working by the law that the energy of the current varies as the square of the current, the energy representing 1hp. would be 27·5 webers in each ohm. It is of course impossible to realise these forces in the working machines. The main object must be to obtain as much of the energy as possible in the useful form of magnetic attraction and repulsion, and as little as may be in the form of heat in the magnets or conductors. It is probably impossible to say what such a current as is here spoken of would cost as obtained from a source such as a voltaic battery consuming zinc. The voltaic cell is unsteady in working, and varies greatly in local action, which is usually a source of much waste. Its resistance also varies. The theoretical expenditure of zinc is, for each weber maintained for ten hours, 175 grains.

When the electro-magnetic engine is furnished with a permanent magnetic field, instead of an electro-magnetic field, there is a certain diminution in the working expense, but it is probably not more than 10 per cent., and is usually more than counterbalanced by the inferiority in power of the permanent field machines, weight for weight, when compared with the electro-magnetic field machines. Reckoning the first cost of the motor in both cases, the electro-magnetic field motor would generally be regarded as the cheaper one throughout. It may be observed that, as work is added, the motor revolves more slowly, and judging by the theories of the inverse electro-motive force exerted by the motor, this results in increased expenditure of zinc in the cell. There is therefore a minimum speed at which electro-motors should revolve.

From the foregoing considerations it would appear that, allowing for all sources of loss, a Bunsen battery, to maintain 1hp. through an efficient motor for ten hours, would cost for materials, oxidants, and excitants about 25s. This amount of work might be done by a common steam engine for one-fifth the amount, allowing for special attendance, the expenditure for coal alone being about 1s. The Daniell cell could probably be made to produce small moving powers at a more economical rate, and it might prove of considerable value in connection with accumulators, but all hope of superseding steam engines by electrical energy derived from zinc must be put aside, even although Joules's calculation of 1s. per hour were practicable. But there are numerous applications of motive energy where steam energy would be impracticable, and in such cases, if not exceeding an expenditure of more than 1,000 foot-pounds, or the thirty-third part of a horse power, an electro-motor might advantageously be used, preferably working from an accumulator.

Local Action.—This source of loss in the working of zinc cells has been referred to above. Any impurities present in a zinc plate, such as traces of iron, lead, or tin, must, when the plate is plunged in an excitant, assume either an electro-positive or an electro-negative condition in relation to the zinc itself, or in relation to other impurities in the plate. From this consideration it is not difficult to perceive that *local* currents, as they are called, would be established from one portion of the zinc to another portion. The real causes at work in local action have not, however, been fully investigated. It is sufficient to know that unprotected zinc is consumed in an excitant by local action, and that the more impure the zinc the more rapid the waste. A remedy for this troublesome defect has long been in use : it consists simply in coating the zinc with a film of mercury, forming what is known as amalgamated zinc. The process itself is simple enough, and may be conducted as follows :—

Amalgamation of the Zinc.—The surface of the zinc should be clean. New zinc plates frequently have an unctuous surface, which may be quickly removed by immersion in a hot solution of caustic soda.

The mercury used should be as pure as possible, or at least free from lead. It may be poured into a shallow vessel, and freely rubbed over the plates after they have been dipped in acidulated water. Dilute sulphuric acid may also be used in the rubber, which may be a pad of cloth or tow. The two metals freely combine, and after a uniform surface is presented the plates should be set up on edge to drain. Very thin zinc plates become saturated with the mercury, and are then very easily broken. Even thick plates must be handled with additional care after being amalgamated. This property of mercury may be turned to useful account in cutting the thick sheet zinc from which the plates are made. A scratch may be scored on either side on the required line of division and mercury run into the grooves, with the aid of a wetted rubber, when the plate may be easily broken across the edge of a table. It should be observed that zinc intended to be curved to a cylindrical form should not be amalgamated previously.

Coercive Force.—A sensible time elapses before soft iron, placed in a magnetic field, or within a current-bearing helix, assumes the magnetic condition. The length of time depends upon two conditions: the hardness of the iron and the intensity of the magnetic field or external circulating current. When the iron is hard, and the magnetic field or current of feeble intensity or strength, the lapse of time is greatest; and conversely, when the iron is soft, in an intense magnetic field or surrounded by a strong current, the lapse of time is small.

Similarly, when the iron is withdrawn from the

magnetic field or current-bearing helix, it does not lose its magnetic properties instantaneously. The rate of decrease is as gradual as the rate of increase. The cause of this is commonly known as *coercive force*. The term "soft iron" is used to denote that kind of iron which exhibits the coercive property to the least degree.

Relative Intensities of Magnetisation of Metals Placed in a Given Magnetic Field.—Soft wrought iron, 32·8; Cast iron, 23·0; Soft steel, 21·6; Hard cast steel, 16·1. (Barlow.)

www.ingramcontent.com/pod-product-compliance
Lightning Source LLC
Chambersburg PA
CBHW021116180426
43192CB00047B/803